Mighty Mentals

Book B

Peter Maher

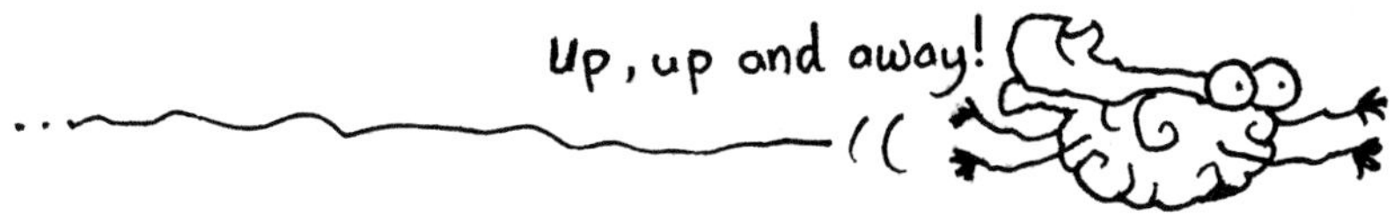

Name ______________________________

Class ______________________________

Contents

Note to teachers and parents

Developing mental arithmetic skills

Automatic response—the ability to quickly recall basic mathematical facts, or to apply a known fact to a slightly modified situation or application—is an essential prerequisite for success in mathematics.

The development of modern technology and its applicability to the classroom in the form of the hand-held calculator has not diminished the significance of the role that mental arithmetic should play in schools. On the contrary, the advent of modern technology has increased the need for accurate mental arithmetic skill development in children. Without relatively accurate estimation prior to calculation, merely accepting at face value a solution on a screen can be fraught with danger. Inadvertently hitting the wrong buttons on a calculator is a commonly occurring fact of life. The result on the screen should never be blindly accepted. When we consider the fact that the great majority of the calculations secondary school students complete are done on the calculator, the validity of this point can be readily perceived.

By the end of primary school, a child should be in the habit of estimating and rounding off prior to using a calculator for computational purposes.

Appropriate estimation is reliant upon a sound knowledge of basic number facts, which should have been developed consistently throughout the primary years.

The better a student's abilities with mental arithmetic, the less time he or she will have to devote to looking up answers on a screen, in a book or on a chart. This point is demonstrated graphically with the game of 'beat the calculator'. Line up your students in pairs and tell them that some maths questions are going to be asked of the children who progressively come to the head of the line. Offer a calculator to one of the two students. Both will want it but only allow one to prevail. The other student will complain about how unfair the game is until the rules are made clear. The child with the calculator must press the appropriate buttons and display the correct answer. Ask the pairs of students questions they should be able to recall relatively quickly, such as 2 x 5 or 7 + 4. Now watch the child with the calculator complain that the game is unfair because the calculator is an encumbrance. Slip in the question 9 x 35 to see who is first with the correct answer. If the student using mental arithmetic is highly competent, 350 – 35 = 315 (10 lots of 35 minus one lot of 35) might still beat the calculator's 9 x 35 = 315.

It is also the case that the faster a student is with his or her basic number facts, the greater is the likelihood of structural computational success. Algorithms, such as long multiplication and division, contain numerous steps and stages. Students with inadequate mental arithmetic skills often take so long to recall a basic number fact that they forget the stage of the algorithm that they are progressing through. For these students the necessary foundation of times tables knowledge is inadequate—it is like trying to build on a foundation of sand. Without the necessary prerequisites of number facts competence, the algorithmic 'house' will tumble down.

It is invariably the case that the broader a student's body of mathematical knowledge, the more he or she will achieve in the subject and the more enjoyment that student will derive from the discipline. The message is clear: regularly practise automatic response games and activities, chants and rounds and all will benefit. Try to teach these number facts in as fun and meaningful a way as possible. Incorporate concrete materials and incentives into your number facts lessons.

Above all, teach for understanding. When teaching the 3 times table, show that the digits in the multiples will always sum to 3, 6 or 9. Show that the answers to the 9 times table always sum to 9. Show that adding on 8 is no harder than adding on 10—just jump back 2. Explain that the metric prefix 'kilo' stands for the number 1000, therefore there have to be 1000 grams in a kilogram and 1000 metres in a kilometre and, heaven forbid, 1000 tons of TNT in a kiloton bomb.

The patterns that make number-facts recall easy are out there and easy enough to find. Make the effort to demonstrate them at every possible opportunity. This book and the other books in this series will show you how.

How to use this book

The challenge of acquiring appropriate automatic response capabilities demands a structured and planned approach in schools. Mental arithmetic skills should be developed in a systematic manner over a period of years. The program should encompass the full range of the mathematics curriculum, covering measurement and space facts, data and chance questions as well as number problems.

The four books in this series assist with the development of mental arithmetic skills in a systematic and logical manner. Each book in the series contains 32 units of work. Each unit contains six sets of questions, which can represent one set per school day with one left over for homework. Alternatively, one day per week could be devoted to the completion of the six sets.

Each exercise in the book focuses on a specific concept, such as odd and even numbers, or digital and analogue time conversions. The exercises have been set out in a logical sequence according to topic, for example multiplying by six or 12 should be taught prior to dividing by six or 12; the concept of length should be taught prior to its application with the concepts of perimeter and area. However, the order in which the exercises are completed can be varied to suit the needs of your students. The exercises are at their most useful when linked to the topic under review in the classroom at the time.

The **Scope and Sequence Chart** on the inside front cover of this book will assist you to plan for the use of this book and to make it as relevant as possible to your students' individual learning capabilities.

At the top of each unit of work there is a **tip** for the students, to help them work through the exercises in that unit. At the bottom of each unit, you will find the relevant state **outcomes**.

The **Handy Maths Facts** on pages 4–5 are included as useful reference material for students.

How Well I Am Doing? on pages 6–7 affords the students the opportunity to record their scores on the unit exercises and to rate their own progress and performance.

A list of **answers** is found in the middle of each book and can be removed and stored if deemed appropriate.

The terms used in this book

The questions presented in this book utilise a wide variety of mathematical terms that are designed to reflect the structure of the discipline and to develop a student's mathematical vocabulary. Knowing, for example, that the terms 'add', 'sum', 'total' and 'altogether' all refer to the process of addition will have obvious benefits when a student has to tackle an addition problem set in context, or when problem solving.

A **Glossary** is provided on page 72 as a ready reference.

Handy Maths Facts

+ or −

	0	1	2	3	4	5	6	7	8	9	10	11	12
0	0	1	2	3	4	5	6	7	8	9	10	11	12
1	1	2	3	4	5	6	7	8	9	10	11	12	13
2	2	3	4	5	6	7	8	9	10	11	12	13	14
3	3	4	5	6	7	8	9	10	11	12	13	14	15
4	4	5	6	7	8	9	10	11	12	13	14	15	16
5	5	6	7	8	9	10	11	12	13	14	15	16	17
6	6	7	8	9	10	11	12	13	14	15	16	17	18
7	7	8	9	10	11	12	13	14	15	16	17	18	19
8	8	9	10	11	12	13	14	15	16	17	18	19	20
9	9	10	11	12	13	14	15	16	17	18	19	20	21
10	10	11	12	13	14	15	16	17	18	19	20	21	22
11	11	12	13	14	15	16	17	18	19	20	21	22	23
12	12	13	14	15	16	17	18	19	20	21	22	23	24

x or ÷

	0	1	2	3	4	5	6	7	8	9	10	11	12
0	0	0	0	0	0	0	0	0	0	0	0	0	0
1	0	1	2	3	4	5	6	7	8	9	10	11	12
2	0	2	4	6	8	10	12	14	16	18	20	22	24
3	0	3	6	9	12	15	18	21	24	27	30	33	36
4	0	4	8	12	16	20	24	28	32	36	40	44	48
5	0	5	10	15	20	25	30	35	40	45	50	55	60
6	0	6	12	18	24	30	36	42	48	54	60	66	72
7	0	7	14	21	28	35	42	49	56	63	70	77	84
8	0	8	16	24	32	40	48	56	64	72	80	88	96
9	0	9	18	27	36	45	54	63	72	81	90	99	108
10	0	10	20	30	40	50	60	70	80	90	100	110	120
11	0	11	22	33	44	55	66	77	88	99	110	121	132
12	0	12	24	36	48	60	72	84	96	108	120	132	144

Length

10 mm = 1 cm

1000 mm = 1 m

100 cm = 1 m

1000 m = 1 km

mm millimetre(s)

cm centimetre(s)

m metre(s)

km kilometre(s)

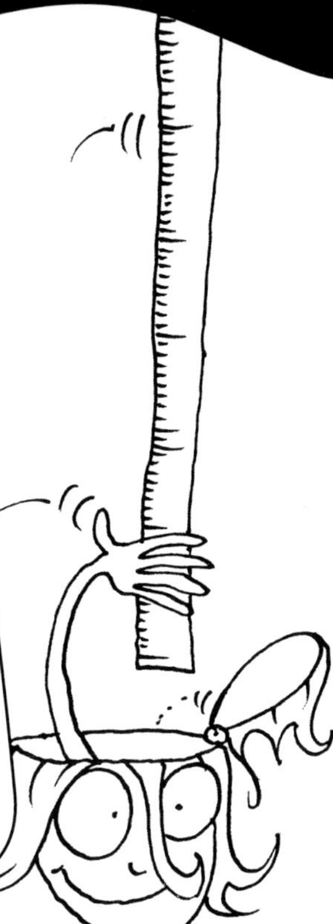

Mass

1000 g = 1 kg

1000 kg = 1 t

g gram(s)

kg kilogram(s)

t tonne(s)

Capacity

1000 mL = 1 L

1000 cm^3 = 1 L

1 cm^3 (water) weighs 1 g

mL millilitre(s)

L litre(s)

g gram(s)

cm^3 cubic centimetres

Time

60 seconds = 1 minute

60 minutes = 1 hour

24 hours = 1 day

7 days = 1 week

365 days = 1 year

366 days = 1 leap year

12 months = 1 year

10 years = 1 decade

100 years = 1 century

How well am I doing?

After you complete each set of work, fill in your score out of 10.
At the end of the six sets, rate yourself in the following way:

S = Some work to do; G = Getting stronger; F = Finalist; or C = Champion!

Unit	Topic	Set A	Set B	Set C	Set D	Set E	Set F	My Rating
1	**Addition:** Bonding to 20							
2	**Addition:** Adding on multiples of 10							
3	**Addition:** Adding on 1–9							
4	**Subtraction:** Subtracting from 20							
5	**Subtraction:** Subtracting multiples of 10							
6	**Subtraction:** Subtracting 1–9							
7	**Doubling and near doubling**							
8	**Halving and near halving**							
9	**Multiplication:** The 6 times and 12 times tables							
10	**Multiplication:** The 7 times and 8 times tables							
11	**Multiplication:** The 9 times and 11 times tables							
12	**Division:** Dividing by 6 and 12							
13	**Division:** Dividing by 7 and 8							
14	**Division:** Dividing by 9 and 11							

Unit	Topic	Set A	Set B	Set C	Set D	Set E	Set F	My Rating
15	**Mixed multiplication and division**							
16	**Odd and even numbers**							
17	**Place value:** Digit recognition							
18	**Rounding off:** Nearest 10, 100 and 1000							
19	**Sequences:** Finding terms and rules							
20	**Number strings**							
21	**Fractions and decimals**							
22	**Length:** Units and conversions							
23	**Perimeter:** Concept and missing sides							
24	**Area:** Concept and application							
25	**Mass:** Units and conversions							
26	**Capacity/volume:** Units and conversions							
27	**Time:** Digital and analogue conversions							
28	**Time:** Unit equivalences							
29	**Money:** Economical ways of making amounts							
30	**2D and 3D:** Properties and directions							
31	**Chance:** Concepts of likelihood							
32	**Revision:** All sorts							

Number and Algebra

Unit 1 Addition: Bonding to 20

Our number system is based on the number 10 because we have 10 fingers. See if you can bond to 10 + 10 (20) quickly!

Set A

1 20 = 5 + □

2 20 = 19 + □

3 20 = 2 + □

4 20 = 16 + □

5 20 = 11 + □

6 20 = 3 + □

7 20 = 18 + □

8 20 = 7 + □

9 20 = 14 + □

10 20 = 0 + □

Set B

1 20 = 12 + 3 + □

2 20 = 5 + 11 + □

3 20 = 2 + 12 + □

4 20 = 3 + 13 + □

5 20 = 3 + 16 + □

6 20 = 13 + 1 + □

7 20 = 4 + 9 + □

8 20 = 5 + 7 + □

9 20 = 9 + 10 + □

10 20 = 7 + 1 + □

Set E

1 Find the sum of 6 and 14 ____

2 Find the sum of 8 and 12 ____

3 Add 3 to the total of 17 and 3 ____

4 What is 9 and 1 and 4 and 6 altogether? ____

5 What is 2 more than the sum of 8 and 12? ____

6 Add 7 to 4 and 16 ____

7 Add 8 to the sum of 9 and 11 ____

8 What is 5 and 15 and 6 altogether? ____

9 What is 6 greater than 9 plus 11? ____

10 Add 9 to the sum of 7 and 13 ____

Set C

1 20 = 2 + 2 + 2 + ☐

2 20 = 3 + 3 + 2 + ☐

3 20 = 1 + 1 + 12 + ☐

4 20 = 5 + 1 + 12 + ☐

5 20 = 4 + 14 + 1 + ☐

6 20 = 3 + 1 + 2 + ☐

7 20 = 2 + 3 + 14 + ☐

8 20 = 7 + 11 + 1 + ☐

9 20 = 0 + 10 + 5 + ☐

10 20 = 4 + 10 + 2 + ☐

Set D

What number is needed to make 20, starting with:

1 **14?** ☐

2 **18?** ☐

3 **15?** ☐

4 **11?** ☐

5 **9?** ☐

6 **8?** ☐

7 **12?** ☐

8 **13?** ☐

9 **20?** ☐

10 **0?** ☐

Set F

1 I need $20 and have saved $8. I still need to save _____.

2 There are 20 books in a set. I have 13 and still need _____ more.

3 We need 20 runs to win the game and have scored 3. We need _____ more runs.

4 I made $20 with a $5 note, two $2 coins and _____ dollar coins.

5 I am missing 20 cents. I found 5 cents. I still need to find _____ cents.

6 I need to do 20 sums for homework. I have completed 9. _____ more sums to go!

7 I must travel a journey of 20 kilometres on my bike. I am at the 4 kilometre mark. I have _____ kilometres to go.

8 I have to iron 20 shirts and have finished 11. _____ shirts to go.

9 My dad wants to lose 20 kilograms on his new diet. He has already lost 3 kilograms. _____ kilograms to go!

10 I must paint 20 metres of fencing. 5 metres have been painted. _____ metres still to paint.

Number and Algebra

Unit 2
Addition: Adding on multiples of 10

Adding on 10, 20, 30 etc. to a number is as simple as adding 1, 2 or 3 etc. to the tens place. The units are not affected.

See how well you can do this with the following questions.

Set A

1 3 + 20 = ☐
2 5 + 40 = ☐
3 7 + 30 = ☐
4 4 + 50 = ☐
5 8 + 70 = ☐
6 6 + 90 = ☐
7 1 + 30 = ☐
8 9 + 80 = ☐
9 2 + 50 = ☐
10 8 + 60 = ☐

Set B

1 80 + 4 = ☐
2 40 + 8 = ☐
3 20 + 2 = ☐
4 60 + 5 = ☐
5 30 + 7 = ☐
6 30 + 1 = ☐
7 50 + 3 = ☐
8 90 + 6 = ☐
9 70 + 9 = ☐
10 80 + 1 = ☐

Set E

1 Find the sum of 23 and 70 ____
2 Add 27 to 50 ____
3 What is the total when 20 is added to 55? ____
4 What is 43 and 30 altogether? ____
5 What is the sum of 60 and 19? ____
6 Add 40 to 46 ____
7 Find the total of 68 and 30 ____
8 What is 70 more than the sum of 3 and 8? ____
9 Is the sum of 36 and 20 more than 55? ____
10 Is the total of 60 and 12 and 10 equal to 92? ____

Set C

1 $5 + 40 + 2 =$ ☐

2 $4 + 20 + 3 =$ ☐

3 $7 + 60 + 1 =$ ☐

4 $2 + 50 + 2 =$ ☐

5 $6 + 70 + 1 =$ ☐

6 $4 + 30 + 4 =$ ☐

7 $1 + 80 + 8 =$ ☐

8 $0 + 70 + 3 =$ ☐

9 $2 + 2 + 2 + 90 =$ ☐

10 $3 + 1 + 3 + 40 =$ ☐

Set D

1 $21 + 70 =$ ☐

2 $43 + 40 =$ ☐

3 $38 + 30 =$ ☐

4 $23 + 50 =$ ☐

5 $15 + 40 =$ ☐

6 $56 + 20 =$ ☐

7 $49 + 50 =$ ☐

8 $45 + 20 =$ ☐

9 $17 + 80 =$ ☐

10 $44 + 30 =$ ☐

Set F

1 I had 27 posters and then was given 20 more. I now have ____ posters.

2 Sam has 40 more posters than the answer above. Sam has ____ posters.

3 Sara has shot 37 netball goals. Her personal best is 20 more than this total. Her PB is ____ goals.

4 I saved $50 and then was given $33. I then had ____.

5 Marie scored 30 more runs than me. I scored 53 runs. She scored ____ runs.

6 I own 11 pairs of shoes. Elisa has 20 more pairs than me. She owns ____ pairs of shoes.

7 In the morning it was 7 degrees. Then it rose 30 degrees. The temperature was then ____ degrees.

8 I weigh 31 kilograms. My dad weighs 40 kilograms more. He weighs ____ kilograms.

9 My team scored 47 points and lost by 30. Our opponents scored ____ points.

10 I scored 46 points for long jump, 50 points less than the winner. The winner scored ____ points.

Number and Algebra

Unit 3 Addition: Adding on 1–9

Remember to start with the largest number first and use strategies; e.g. when adding on 7, 8 or 9 add on 10 and jump back 1, 2 or 3.

Adding on 1, 2, 3, 4, 5 or 6 will become easier with practice.

Set A

1 17 + 2 = ☐

2 15 + 1 = ☐

3 11 + 3 = ☐

4 16 + 2 = ☐

5 13 + 3 = ☐

6 1 + 12 = ☐

7 2 + 19 = ☐

8 3 + 16 = ☐

9 2 + 15 = ☐

10 3 + 16 = ☐

Set B

1 13 + 4 = ☐

2 18 + 6 = ☐

3 12 + 5 = ☐

4 17 + 4 = ☐

5 14 + 5 = ☐

6 5 + 15 = ☐

7 4 + 11 = ☐

8 5 + 19 = ☐

9 4 + 12 = ☐

10 6 + 15 = ☐

Set E

1 Add 9 to 27 _____

2 Add 4 to 56 _____

3 What is the sum of 34 and 7? _____

4 What is the sum of 72 and 3? _____

5 What is 48 and 5 altogether? _____

6 What is 48 and 6 altogether? _____

7 Find the total of 6 and 32 _____

8 Find the total of 36 and 9 _____

9 Is the sum of 75 and 7 more than 82? _____

10 Is the sum of 23 and 8 equal to 31? _____

Set C

1 15 + 7 = ☐
2 19 + 9 = ☐
3 17 + 8 = ☐
4 11 + 9 = ☐
5 19 + 9 = ☐
6 8 + 13 = ☐
7 9 + 15 = ☐
8 7 + 16 = ☐
9 8 + 18 = ☐
10 7 + 19 = ☐

Set D

1 46 + 7 = ☐
2 83 + 4 = ☐
3 52 + 5 = ☐
4 92 + 6 = ☐
5 74 + 9 = ☐
6 9 + 67 = ☐
7 3 + 48 = ☐
8 6 + 55 = ☐
9 2 + 97 = ☐
10 5 + 81 = ☐

Set F

1 In the morning it was 6 degrees. It rose 25 more degrees in the afternoon. It was then _____ degrees.

2 It was 13 degrees in Hobart. It was 8 degrees hotter in Adelaide. In Adelaide it was _____ degrees.

3 My team scored 66 points but lost by 8 points. The winning team scored _____ points.

4 At half time in the netball the Phoenix were 37 goals. They scored 6 goals in the third quarter. At three-quarter time they had scored _____ goals.

5 I owned 39 lollies and then was given 5 more. I now have _____ lollies.

6 My brother has 4 more lollies than this total. How many does he have? _____

7 I own 17 CDs and was given 7 more for my birthday. I now own _____ CDs.

8 I have ridden 92 kilometres on my trail bike. My brother has ridden 6 kilometres more than me. He has ridden _____ kilometres on his bike.

9 I am 9 years old. My sister is 3 years older than me. She is _____ years old.

10 There are 23 trees in my garden and 8 in next door's garden. The 2 houses have _____ trees altogether.

NSW: Number NS2.2; Qld: Number concepts N3.2;
Vic: Mental computation and estimation 3.2; WA: Number N3.3; SA: Number 2.8

Number and Algebra

Unit 4 Subtraction: Subtracting from 20

If you can bond to 20, taking away from 20 will be very easy. Just remember that subtraction is simply the opposite of addition.

Set A

1 20 – 10 = ☐

2 20 – 18 = ☐

3 20 – 11 = ☐

4 20 – 17 = ☐

5 20 – 16 = ☐

6 20 – 13 = ☐

7 20 – 12 = ☐

8 20 – 15 = ☐

9 20 – 19 = ☐

10 20 – 14 = ☐

Set B

1 20 – 3 = ☐

2 20 – 7 = ☐

3 20 – 5 = ☐

4 20 – 9 = ☐

5 20 – 1 = ☐

6 20 – 8 = ☐

7 20 – 0 = ☐

8 20 – 4 = ☐

9 20 – 6 = ☐

10 20 – 2 = ☐

Set E

1 How many times can you take away 5 from 20? ____

2 How many times can 2 be taken away from 20? ____

3 How many times can you take away 1 from 20? ____

4 What is left over if you take 4, then 6 away from 20? ____

5 What is left over after taking away 3, twice, from 20? ____

6 Can you take 7 away from 20, 3 times? ____

7 Remove 8 from 20 and add on 2 ____

8 Remove 5 from 20 and double the answer ____

9 Remove 2 from 20, 4 times ____

10 Remove 16 and then 2 and then 1 from 20 ____

Set C

How many more do you need to take away from 20 to reach 0?

1	4	☐
2	7	☐
3	1	☐
4	5	☐
5	0	☐
6	3	☐
7	6	☐
8	8	☐
9	2	☐
10	9	☐

Set D

1 **20 – 2 – 7 =** ☐

2 **20 – 8 – 1 =** ☐

3 **20 – 0 – 0 – 9 =** ☐

4 **20 – 4 – 4 =** ☐

5 **20 – 5 – 5 =** ☐

6 **20 – 3 – 3 – 3 =** ☐

7 **20 – 5 – 3 – 2 =** ☐

8 **20 – 1 – 1 – 1 =** ☐

9 **20 – 8 – 1 – 4 =** ☐

10 **20 – 7 – 1 – 2 – 3 =** ☐

Set F

1 I had 20 cents and spent 15 cents. I had ____ cents left.

2 It was 20 degrees and then fell 11 degrees. It was then ____ degrees.

3 My birthday is on the 20th of June. My friend was born on the 12th of June. I am ____ days older than she is.

4 I can't find 4 of my 20 pairs of socks. I can only find ____ pairs of socks.

5 My team scored 20 goals and I scored 7 of them. The rest of the team scored ____ goals between them.

6 I had to rule a 20-centimetre line, but my line was 8 centimetres short. My line was ____ centimetres long.

7 My dad's birthday is the 20th of May. Today is the 8th of May. He still must wait ____ more days.

8 My dog weighs 20 kilograms and my cat weighs 9 kilograms. My dog is ____ kilograms heavier than my cat

9 Of the 20 t-shirts I own, 6 are too small. Only ____ fit me.

10 3 of my 20 video games are rally-driving games. ____ are not rally-driving games.

Number and Algebra

Unit 5 Subtraction: Subtracting multiples of 10

Just like when we add on 10 or 20 or 30, when 10 or 20 or 30 is taken away from a number, we simply need to change the tens place. The units are not affected.

Set A

1 37 – 30 = ☐
2 28 – 20 = ☐
3 29 – 20 = ☐
4 32 – 30 = ☐
5 94 – 90 = ☐
6 35 – 30 = ☐
7 18 – 10 = ☐
8 27 – 20 = ☐
9 31 – 30 = ☐
10 33 – 30 = ☐

Set B

1 46 – 20 = ☐
2 37 – 10 = ☐
3 74 – 20 = ☐
4 51 – 30 = ☐
5 88 – 30 = ☐
6 24 – 10 = ☐
7 95 – 20 = ☐
8 66 – 30 = ☐
9 93 – 20 = ☐
10 59 – 30 = ☐

Set E

1 Take 20 from 37 ____
2 Take 30 from 72 ____
3 Subtract 20 from 84 ____
4 Subtract 30 from 66 ____
5 Reduce 71 by 20 ____
6 Reduce 94 by 30 ____
7 Find the difference between 64 and 20 ____
8 Find the difference between 72 and 30 ____
9 Remove 20 from 87 ____
10 Remove 30 from 91 ____

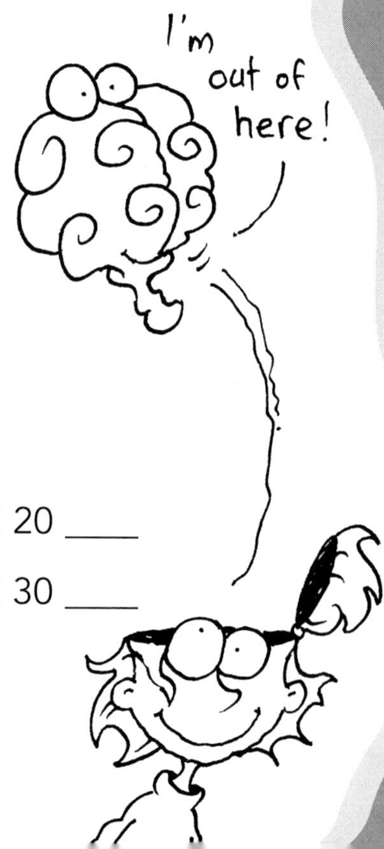

Set C

1 134 – 20 = ☐
2 177 – 20 = ☐
3 144 – 30 = ☐
4 138 – 30 = ☐
5 165 – 30 = ☐
6 188 – 20 = ☐
7 161 – 20 = ☐
8 172 – 30 = ☐
9 191 – 20 = ☐
10 178 – 30 = ☐

Set D

1 54 – 20 – 20 = ☐
2 37 – 20 – 10 = ☐
3 86 – 20 – 20 = ☐
4 87 – 20 – 10 = ☐
5 44 – 20 – 20 = ☐
6 57 – 10 – 20 = ☐
7 88 – 10 – 20 = ☐
8 61 – 30 – 10 = ☐
9 94 – 20 – 20 – 20 = ☐
10 82 – 10 – 20 – 20 = ☐

Set F

1 I had 35 cents and spent 20 cents. I had _____ cents left.

2 I saved $47 and then bought a game costing $30. I had _____ left.

3 20 of our 67 Year 4 students were sick. _____ were at school.

4 The car race was 74 km long . After 30 km the drivers had ___ km to go.

5 I lost 20 of my 59 stickers. I had _____ stickers left.

6 We scored 57 goals and won by 30. Our opponents scored _____ goals.

7 Southside scored 83 runs and beat us by 20. We scored _____ runs.

8 In Darwin it was 34 degrees, 20 degrees more than in Hobart. In Hobart it was _____ degrees.

9 Grandma is 63 and is 30 years older than Mum. Mum is _____ years old.

10 On the test I scored 89 and beat the average by 20.The average score was _____.

Number and Algebra

Unit 6 Subtraction: Subtracting 1–9

Subtracting 1, 2 or 3 is easy.

Subtracting 9 and 8 is like subtracting 10 and adding on 1 or 2.

Subtracting 4, 5, 6 and 7 will become easy with practice.

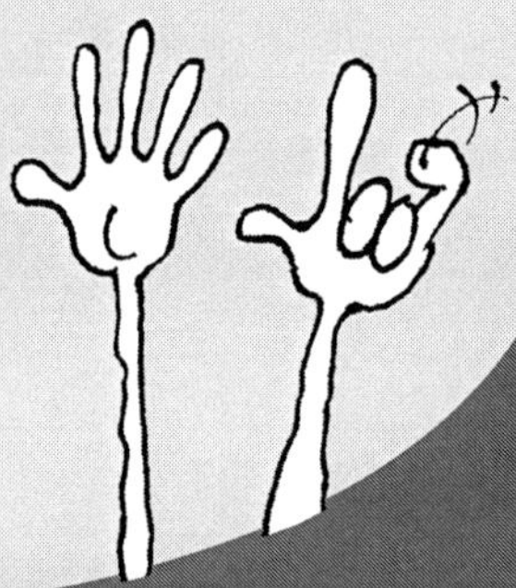

Set A

1. 8 – 2 = ☐
2. 5 – 3 = ☐
3. 9 – 2 = ☐
4. 9 – 1 = ☐
5. 7 – 2 = ☐
6. 14 – 2 = ☐
7. 17 – 3 = ☐
8. 23 – 2 = ☐
9. 37 – 3 = ☐
10. 45 – 1 = ☐

Set B

1. 11 – 8 = ☐
2. 24 – 9 = ☐
3. 33 – 8 = ☐
4. 72 – 9 = ☐
5. 65 – 9 = ☐
6. 94 – 8 = ☐
7. 44 – 9 = ☐
8. 55 – 8 = ☐
9. 70 – 8 = ☐
10. 74 – 9 = ☐

Set E

1. Take 5 away from 43 ___
2. Take 9 away from 71 ___
3. Reduce 55 by 4 ___
4. Reduce 33 by 7 ___
5. Subtract 2 from 21 ___
6. Subtract 3 from 82 ___
7. What is 17 minus 8? ___
8. What is 64 minus 6? ___
9. What is the difference between 82 and 4? ___
10. What is the difference between 9 and 77? ___

Set C

1 29 – 4 = ☐
2 73 – 5 = ☐
3 33 – 6 = ☐
4 82 – 7 = ☐
5 21 – 4 = ☐
6 82 – 5 = ☐
7 92 – 6 = ☐
8 28 – 7 = ☐
9 47 – 5 = ☐
10 81 – 7 = ☐

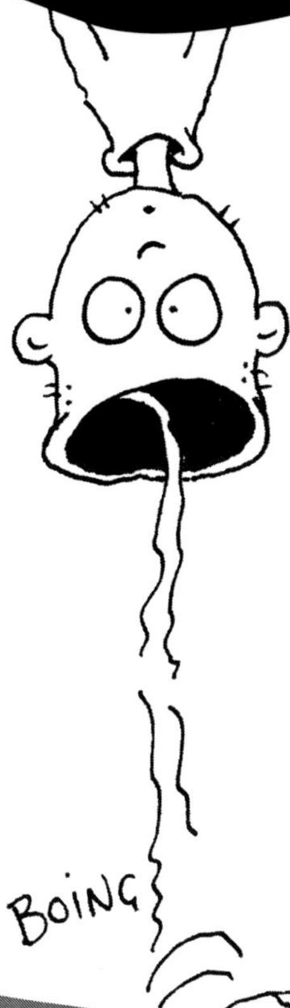

Set D

1 44 – 8 = ☐
2 27 – 3 = ☐
3 54 – 6 = ☐
4 36 – 2 = ☐
5 91 – 4 = ☐
6 78 – 5 = ☐
7 45 – 7 = ☐
8 80 – 1 = ☐
9 77 – 9 = ☐
10 28 – 5 = ☐

Set F

I shared my 100 lollies with my friends.

1 I gave 5 to Sara and had ____ left.
2 I gave 6 lollies to Robert and then had ____ left.
3 I gave 2 lollies to Tran and then had ____ left.
4 I gave 8 lollies to Claire and then had ____ left.
5 I gave 7 lollies to Caitlin and then had ____ left.
6 I gave 9 lollies to Aiko and then had ____ left.
7 I gave 5 to Andrew and then had ____ left.
8 I gave 9 to Joel and then had ____ left.
9 I gave 6 to Nicholas and then had ____ left.
10 I gave 7 to Jamil and kept the rest. I had ____ lollies for myself.

Number and Algebra

Unit 7 Doubling and near doubling

Knowing how to add a number onto itself (doubling) is a useful skill. If you know the 2 times table, you can double numbers. Near doubling just needs a little extra step. Double a 2-digit number by doubling the tens and doubling the units and adding them together.

Set A

Double the following numbers.

1	7	
2	3	
3	11	
4	9	
5	5	
6	8	
7	4	
8	6	
9	10	
10	12	

Set B

Double the answers to the following questions.

1	5 + 3	
2	5 + 5	
3	4 + 3	
4	6 + 2	
5	8 + 4	
6	4 + 8	
7	6 + 3	
8	2 + 8	
9	5 + 6	
10	3 + 8	

Set E

1	6 + 7 =	
2	8 + 7 =	
3	11 + 12 =	
4	15 + 14 =	
5	22 + 21 =	
6	33 + 32 =	
7	25 + 27 =	
8	44 + 42 =	
9	32 + 31 =	
10	19 + 22 =	

Set C

Double the following numbers.

1 24 ☐

2 33 ☐

3 44 ☐

4 23 ☐

5 21 ☐

6 31 ☐

7 22 ☐

8 41 ☐

9 34 ☐

10 43 ☐

Set D

Double the following numbers.

1 25 ☐

2 45 ☐

3 15 ☐

4 35 ☐

5 16 ☐

6 27 ☐

7 36 ☐

8 48 ☐

9 29 ☐

10 37 ☐

Set F

1 I saved $34 and was given $34 for my birthday. I then had ____.

2 It was 16 degrees in the morning and doubled in temperature in the afternoon. It was then ____ degrees.

3 We kicked 13 goals but our opponents doubled this score. They kicked ____ goals.

4 Dad has 18 pairs of shoes and Mum has twice as many. She has ____ pairs of shoes.

5 We doubled Southside's score. They scored 34 runs. We scored ____ runs.

6 I weigh 27 kg. My big brother weighs twice this total. He weighs ____ kg.

7 There are 39 children in Year 4 and twice this number in Year 6. In Year 6 there are ____ children.

8 In 2004 I had 52 stickers. Now I have twice this total. I now have _____ stickers.

9 In 2003 oil was $26 a barrel. In 2004 it doubled in price to _____ a barrel.

10 Start with 5. Double the number again and again and again. The number is now ____.

Number and Algebra

Unit 8 Halving and near halving

If you can double a number, then you can halve it as well because one is the opposite of the other. It's like saying that if you can add, you must be able to subtract as well, because they are opposites too.

Set A

Halve the following numbers.

1	**12**	
2	**22**	
3	**46**	
4	**82**	
5	**66**	
6	**48**	
7	**62**	
8	**88**	
9	**44**	
10	**68**	

Set B

Halve the following numbers.

1	**30**	
2	**80**	
3	**40**	
4	**70**	
5	**100**	
6	**90**	
7	**60**	
8	**50**	
9	**110**	
10	**120**	

Set E

1 Find half of 46 _____

2 Find half of 122 _____

3 What is half of 420? _____

4 What is half of 248? _____

5 44 is half of _____

6 333 is half of _____

7 $2\frac{1}{2}$ is half of _____

8 $5\frac{1}{2}$ is half of _____

9 $20\frac{1}{2}$ is half of _____

10 $50\frac{1}{2}$ is half of _____

Set C

Halve the following numbers.

1. 242
2. 288
3. 442
4. 640
5. 888
6. 266
7. 480
8. 688
9. 808
10. 608

Set D

Halve the following numbers.

1. 3
2. 7
3. 11
4. 15
5. 21
6. 29
7. 31
8. 37
9. 43
10. 49

Set F

1 I kicked 42 goals for the year, half of what the full forward kicked. He kicked ____ goals.

2 Only half of our 66 members turned up for the meeting. There were ____ members there.

3 Half of our 30 students have colds. ____ students have colds.

4 I am on page 52 of my book and am halfway through it. It contains _____ pages.

5 Half of a number is $3\frac{1}{2}$. The number is ____.

6 Half of my 84 stickers are swaps. I have ____ swaps.

7 I spent half of my $11 pocket money on lollies. I have _______ left.

8 Half of my $8 is in 50 cent coins. I have ____ 50 cent coins.

9 I watched TV for half of the time between 9 and 11 o'clock. I watched for ____ minutes.

10 I could only run for half of the 2 km race. I ran ______ metres.

Number and Algebra

Unit 9
Multiplication: The 6 times and 12 times tables

Numbers in the 6 times table are even and are all in the 3 times table, so their digits must sum to 3, 6 or 9. Numbers in the 12 times table must be in the 3 times table and the 4 times table.

Set A

1 7 x 6 = ☐

2 2 x 6 = ☐

3 11 x 6 = ☐

4 5 x 6 = ☐

5 9 x 6 = ☐

6 12 x 6 = ☐

7 3 x 6 = ☐

8 8 x 6 = ☐

9 4 x 6 = ☐

10 6 x 6 = ☐

Set B

1 12 x 4 = ☐

2 12 x 8 = ☐

3 12 x 7 = ☐

4 12 x 12 = ☐

5 12 x 3 = ☐

6 12 x 6 = ☐

7 12 x 5 = ☐

8 12 x 9 = ☐

9 12 x 2 = ☐

10 12 x 10 = ☐

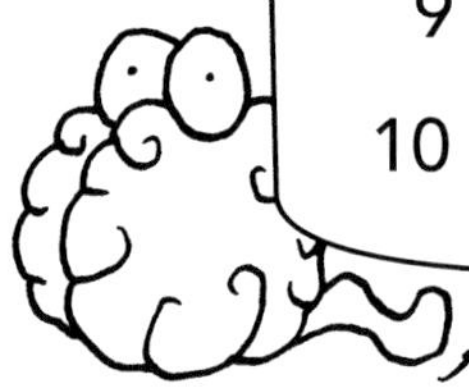

Set E

1 Find the product of 4 and 6 _____

2 Find the product of 4 and 12 _____

3 Why is the answer to Question 2 twice the answer to Question 1? ______________________________

4 Find the 3rd multiple of 6 _____

5 Find the 10th multiple of 12 _____

6 Is 72 the 8th multiple of 12? _____

7 Multiply 8 by 6 _____

8 Multiply 7 by 12 and add 4 _____

9 Take 2 from the product of 6 and 6 _____

10 Take 8 from the product of 12 and 12 _____

Set C

1 $4 \times 6 + 2 =$ ☐

2 $3 \times 12 + 3 =$ ☐

3 $10 \times 6 + 7 =$ ☐

4 $5 \times 6 + 10 =$ ☐

5 $6 \times 12 + 8 =$ ☐

6 $8 \times 6 - 2 =$ ☐

7 $2 \times 12 - 6 =$ ☐

8 $3 \times 6 - 4 =$ ☐

9 $10 \times 12 - 5 =$ ☐

10 $7 \times 6 - 3 =$ ☐

Set D

Write 6 and/ or 12 if these numbers are in the 6 or 12 times table and 'neither' if they are not.

1 **24** ____

2 **30** ____

3 **22** ____

4 **60** ____

5 **42** ____

6 **50** ____

7 **36** ____

8 **48** ____

9 **54** ____

10 **72** ____

Set F

1 How many sides are on 7 hexagons? ____

2 How many eggs in 4 dozen? ____

3 How many months old are you when you turn 10? ____

4 Triple the number of signs of the zodiac. ____

5 There are 6 periods in a day at school. In a week there are ____ periods at school.

6 How many legs are on 11 bees? ____

7 6 babies born at once are called sextuplets. 9 sextuplets = ____ babies.

8 12 inches made a foot. 5 feet = ____ inches.

9 In Australia's innings, 7 sixes were scored. ____ runs were scored in sixes.

10 In football, a goal equals 6 points. 4 goals equal ____ points.

NSW: Number NS2.3; Qld: Number concepts N3.3;
Vic: Mental computation and estimation 3.1; WA: Number N3.3; SA: Number 2.8

Number and Algebra

Unit 10
Multiplication: The 7 times and 8 times tables

For numbers in the 7 times table, if you double the ones and subtract the tens, the answer will be 0, 7 or 14.
So, 4 x 7 = 28:
double 8 = 16,
16 - 2 = 14.

Every second number in the 4 times table is also in the 8 times table.

Set A

1 5 x 7 ☐
2 7 x 7 ☐
3 11 x 7 ☐
4 4 x 7 ☐
5 8 x 7 ☐
6 12 x 7 ☐
7 6 x 7 ☐
8 10 x 7 ☐
9 3 x 7 ☐
10 9 x 7 ☐

Set B

1 11 x 8 ☐
2 5 x 8 ☐
3 3 x 8 ☐
4 12 x 8 ☐
5 6 x 8 ☐
6 9 x 8 ☐
7 10 x 8 ☐
8 7 x 8 ☐
9 4 x 8 ☐
10 8 x 8 ☐

Set E

1 Multiply 7 by itself ____
2 Multiply 8 by 4 ____
3 Find the product of 12 and 7 ____
4 Find the product of 6 and 8 ____
5 Multiply 8 by itself ____
6 How many 7s make 63? ____
7 How many 8s make 96? ____
8 What number is in both the 7 and 8 times table? ____
9 Which numbers in the 20s are in either the 7 or the 8 times tables? ____________
10 Find the 11th multiple of 7 ____

Set C

1 $3 \times 7 + 2 =$ ☐

2 $4 \times 8 + 10 =$ ☐

3 $11 \times 7 + 5 =$ ☐

4 $5 \times 8 + 11 =$ ☐

5 $10 \times 7 + 12 =$ ☐

6 $8 \times 8 - 4 =$ ☐

7 $7 \times 7 - 6 =$ ☐

8 $6 \times 8 - 8 =$ ☐

9 $12 \times 7 - 6 =$ ☐

10 $11 \times 8 - 7 =$ ☐

Set D

Write '7' or '8' if the number is in one of those tables and 'neither' if it is not in either table.

1 63 ____________

2 65 ____________

3 28 ____________

4 32 ____________

5 56 ____________

6 71 ____________

7 35 ____________

8 48 ____________

9 82 ____________

10 21 ____________

Set F

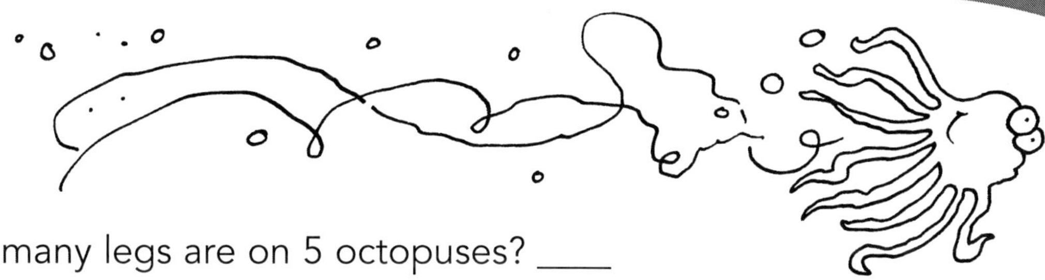

1 How many legs are on 5 octopuses? ____

2 Septuplets are 7 babies born together. 11 septuplets = ____ babies.

3 How many legs are on 3 spiders? ____

4 How many sides are on 8 octagons? ____

5 A heptagon has 7 sides. 7 heptagons = ____ sides.

6 I am 9 and my Grandpa is 8 times as old as me. He is ____ years old.

7 The heptathlon contains 7 events. 6 heptathlons = ____ events.

8 I have 8 five-cent coins. These are worth ____ cents altogether.

9 My 7 ten cent coins are worth ____ cents altogether.

10 I am 9 and my Grandma is 7 times as old as me. She is ____ years old.

Number and Algebra

Unit 11
Multiplication: The 9 times and 11 times tables

Numbers in the 9 times table have their digits summing to 9 or 18. So for, 27: 2 + 7 = 9.

Numbers in the 11 times table have their outside equalling their inside when compared. So, 66: 6 = 6; 121: 1 + 1 = 2.

Set A

1 7 x 9 =

2 4 x 9 =

3 11 x 9 =

4 5 x 9 =

5 8 x 9 =

6 12 x 9 =

7 3 x 9 =

8 6 x 9 =

9 10 x 9 =

10 9 x 9 =

Set B

1 5 x 11 =

2 8 x 11 =

3 2 x 11 =

4 11 x 11 =

5 7 x 11 =

6 6 x 11 =

7 4 x 11 =

8 10 x 11 =

9 9 x 11 =

10 12 x 11 =

Set E

1 Find the product of 3 and 11 ___

2 Find the product of 7 and 9 ___

3 Multiply 8 by 11 ___

4 Multiply 4 by 9 ___

5 What is the 8th multiple of 9? ___

6 What is the 2nd multiple of 11? ___

7 How many elevens make 110? ___

8 How many nines make 108? ___

9 Add 5 to the product of 5 and 9 ___

10 Reduce the product of 11 and 4 by 10 ___

Set C

1 $4 \times 9 + 10 =$

2 $5 \times 11 + 5 =$

3 $8 \times 9 + 3 =$

4 $9 \times 11 + 2 =$

5 $9 \times 9 + 4 =$

6 $7 \times 11 - 6 =$

7 $5 \times 9 - 10 =$

8 $10 \times 11 - 9 =$

9 $12 \times 9 - 8 =$

10 $3 \times 11 - 4 =$

Set D

Write '9' or '11' if the number is in the 9 or 11 times table and 'neither' if the number is not in either table.

1 **63** ________

2 **88** ________

3 **70** ________

4 **44** ________

5 **99** ________

6 **18** ________

7 **121** ________

8 **111** ________

9 **108** ________

10 **37** ________

Set F

1 My piggy bank contains nine 10 cent coins worth ____ cents altogether.

2 My sister collects $5 notes and has 11, worth ______ altogether.

3 A nonagon has 9 sides. 3 nonagons have ____ sides altogether.

4 A hendecagon has 11 sides. 7 hendecagons contain ____ sides altogether.

5 There are 11 players in a soccer team. 8 teams contain ____ players.

6 I live 9 km from school. In a week I travel ____ km to and from school.

7 There are 11 girls in each of our six Year 4 classes. In Year 4 there are ____ girls altogether.

8 My 9 cousins each have 3 pets. My cousins own ____ pets altogether.

9 I made up 11 lolly bags, each containing 10 lollies. I used ______ lollies altogether.

10 I am 9 and my great-uncle is 11 times as old as me. He is ____ years old.

Number and Algebra

Unit 12
Division: Dividing by 6 and 12

Dividing means 'to break up into equal pieces'. Dividing by 6 or 12 means to split something into 6 or 12 equal pieces. Because division is the opposite of multiplication, if you know your 6 and 12 times tables, you can divide by 6 and 12 too!

Set A

Write '6' or '12' if the number can be divided by 6 or 12 without a remainder. Write 'no' if the number can't be divided by either one.

1 **24** ____
2 **42** ____
3 **60** ____
4 **54** ____
5 **28** ____
6 **72** ____
7 **20** ____
8 **30** ____
9 **66** ____
10 **64** ____

Set B

Divide the following numbers by 6.

1	**24**	☐
2	**36**	☐
3	**66**	☐
4	**48**	☐
5	**12**	☐
6	**30**	☐
7	**72**	☐
8	**42**	☐
9	**6**	☐
10	**54**	☐

Set E

What remainder is left when the following numbers are divided by 12?

1 17 ___
2 21 ___
3 34 ___
4 39 ___
5 47 ___
6 56 ___
7 60 ___
8 76 ___
9 89 ___
10 100 ___

Set C

Which numbers have been divided by 12 to produce the following answers?

1	7	
2	11	
3	2	
4	9	
5	0	
6	5	
7	10	
8	8	
9	4	
10	6	

Set D

What remainder is left when the following numbers are divided by 6?

1	21	
2	33	
3	74	
4	55	
5	9	
6	31	
7	70	
8	49	
9	50	
10	17	

Set F

1 How many hexagons equal 72 sides? ____

2 132 loaves of bread equal ____ dozen loaves.

3 60 months are equal to ____ years.

4 24 legs can be found on ____ insects.

5 If you have lived for 96 months you would be ____ years old.

6 12 inches equalled a foot in the olden days. 84 inches = ____ feet.

7 54 points in football equal ____ goals.

8 67 eggs equal ____ dozen with ____ left over.

9 66 runs in sixes equal ____ 6s in cricket.

10 I bowled 18 balls in the cricket match. This equals ____ overs.

Number and Algebra

Unit 13 Division: Dividing by 7 and 8

When we divide by 7 or 8, it is the opposite of multiplying by 7 or 8. So when **21** is split into **7** equal parts, we can find the answer by counting in groups of **7** until we reach **21: 7, 14, 21 = 3** groups. Thus, **21** split into **7** equal parts equals **3.**

Set A

Divide the following numbers by 7.

1	**84**	
2	**49**	
3	**70**	
4	**21**	
5	**28**	
6	**77**	
7	**56**	
8	**14**	
9	**63**	
10	**42**	

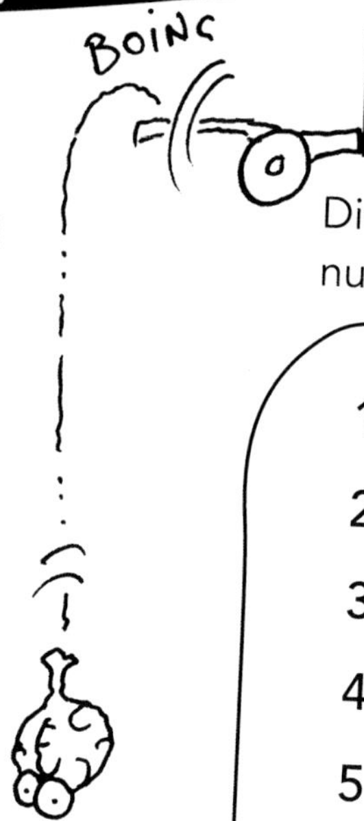

Set B

Divide the following numbers by 8.

1	**80**	
2	**24**	
3	**72**	
4	**32**	
5	**96**	
6	**48**	
7	**40**	
8	**88**	
9	**64**	
10	**56**	

Set E

1. Find the quotient of 7 and 42 ___
2. Find the quotient of 88 and 8 ___
3. Divide 63 by 7 ___
4. Divide 96 by 8 ___
5. How many times can 7 be subtracted from 28? ___
6. How many times can 8 be subtracted from 40? ___
7. What is the remainder when 50 is divided by 7? ___
8. What is the remainder when 50 is divided by 8? ___
9. Divide 56 by 7 and then by 8 ___
10. Split 84 into 7 equal parts. How many in each part? ___

Set C

What is the remainder when the following numbers are divided by 7?

1	9	
2	20	
3	28	
4	31	
5	44	
6	54	
7	60	
8	77	
9	88	
10	90	

Set D

What is the remainder when the following numbers are divided by 8?

1	17	
2	31	
3	35	
4	45	
5	48	
6	62	
7	74	
8	88	
9	92	
10	102	

Set F

1 How many spiders equal 24 legs? ____

2 How many heptagons have 49 sides? ____

3 28 events equal ____ heptathlons.

4 88 sides equal ____ octagons.

5 40 legs are on ____ octopuses.

6 48 children are in teams of 8. There are ____ teams.

7 63 days equal ____ weeks.

8 I have 64 stickers in packets of 8. I have ____ packets.

9 An octave is an 8-line poem. 96 lines are in ____ octaves.

10 84 lollies were put into bags each containing 7 lollies. There were ____ lolly bags.

Number and Algebra

Unit 14
Division: Dividing by 9 and 11

Remember that dividing by 9 and 11 is the exact opposite of multiplying by 9 and 11. If 3 groups of 9 equal 27, then 27 divided by 9 must equal 3. If 8 groups of 11 equal 88, then 88 divided by 11 must equal 8.

Set A

Divide these numbers by 9.

1	99	
2	72	
3	54	
4	36	
5	108	
6	18	
7	81	
8	63	
9	45	
10	90	

Set B

Divide these numbers by 11.

1	55	
2	99	
3	66	
4	132	
5	33	
6	88	
7	44	
8	22	
9	121	
10	110	

Set E

1. Divide 45 by 9 ___
2. Divide 88 by 11 ___
3. How many nines make 27? ___
4. How many elevens make 132? ___
5. What is the quotient of 44 and 11? ___
6. What is the quotient of 72 and 9? ___
7. How many times can 9 be subtracted from 108? ___
8. How many times can 11 be subtracted from 66? ___
9. What is the remainder when 50 is divided by 9? ___
10. What is the remainder when 103 is divided by 11? ___

Set C

What is the remainder when the following numbers are divided by 9?

1	**14**	
2	**26**	
3	**28**	
4	**42**	
5	**45**	
6	**59**	
7	**67**	
8	**79**	
9	**101**	
10	**108**	

Set D

What is the remainder when the following numbers are divided by 11?

1	**15**	
2	**32**	
3	**36**	
4	**52**	
5	**66**	
6	**78**	
7	**93**	
8	**99**	
9	**106**	
10	**123**	

Set F

1 77 cricketers can be put into ____ teams of 11 players.

2 132 soccer players can be put into ____ teams of 11.

3 72 sides are on ____ nonagons.

4 9 buttons are on a shirt. 54 buttons are on ____ shirts.

5 11 rainy days per month. 66 rainy days in ____ months.

6 9 strawberries in a punnet. 81 strawberries in ____ punnets.

7 I live 9 km from school. I travel 36 km to and from school in ____ days.

8 I live 11 km from the city. I travel 88 km when I travel to and from the city ____ times.

9 There are 9 pages in each chapter. 45 pages are in ____ chapters.

10 There are 11 girls in each class. There are 99 girls at school. We have ____ classes at school.

NSW: Number NS2.3; Qld: Number concepts N3.3;
Vic: Mental computation and estimation 3.1; WA: Number N3.3; SA: Number 2.8

Number and Algebra

Unit 15 Mixed multiplication and division

See how well you can remember all the tables and apply them to both multiplication and its opposite, division.

Set A

1 $7 \times 4 =$ ☐

2 $11 \times 8 =$ ☐

3 $5 \times 9 =$ ☐

4 $12 \times 6 =$ ☐

5 $10 \times 3 =$ ☐

6 $2 \times 1 =$ ☐

7 $9 \times 7 =$ ☐

8 $4 \times 5 =$ ☐

9 $8 \times 6 =$ ☐

10 $3 \times 12 =$ ☐

Set B

1 $54 \div 9 =$ ☐

2 $77 \div 11 =$ ☐

3 $63 \div 7 =$ ☐

4 $45 \div 5 =$ ☐

5 $64 \div 8 =$ ☐

6 $14 \div 2 =$ ☐

7 $24 \div 4 =$ ☐

8 $110 \div 10 =$ ☐

9 $84 \div 12 =$ ☐

10 $27 \div 3 =$ ☐

Set E

1 Divide 48 by 4 and add 2 ___

2 Multiply 5 by 6 and add 7 ___

3 Find the quotient of 132 and 12 ___

4 Find the product of 8 and 3 ___

5 What is the remainder when 62 is divided by 10? ___

6 Divide 100 by 10 and then by 5 ___

7 Triple 12 and add 4 ___

8 Divide 60 by 10 and then by 2 ___

9 Multiply 3 by 3 and then by 8 ___

10 Triple 4 and then double this number ___

Answers

Unit 1 Addition: Bonding to 20

	Set A	Set B	Set C	Set D	Set E	Set F
1	15	5	14	6	20	$12
2	1	4	12	2	20	7
3	18	6	6	5	23	17
4	4	4	2	9	20	11
5	9	1	1	11	22	15
6	17	6	14	12	27	11
7	2	7	1	8	28	16
8	13	8	1	7	26	9
9	6	1	5	0	26	17
10	20	12	4	20	29	15

Unit 2 Addition: Adding on multiples of 10

	Set A	Set B	Set C	Set D	Set E	Set F
1	23	84	47	91	93	47
2	45	48	27	83	77	87
3	37	22	68	68	75	57
4	54	65	54	73	73	$83
5	78	37	77	55	79	83
6	96	31	38	76	86	31
7	31	53	89	99	98	37
8	89	96	73	65	81	71
9	52	79	96	97	yes	77
10	68	81	47	74	no	96

Unit 3 Addition: Adding on 1–9

	Set A	Set B	Set C	Set D	Set E	Set F
1	19	17	22	53	36	31
2	16	24	28	87	60	21
3	14	17	25	57	41	74
4	18	21	20	98	75	43
5	16	19	28	83	53	44
6	13	20	21	76	54	48
7	21	15	24	51	38	24
8	19	24	23	61	45	98
9	17	16	26	99	no	12
10	19	21	26	86	yes	31

Unit 4 Subtraction: Subtracting from 20

	Set A	Set B	Set C	Set D	Set E	Set F
1	10	17	16	11	4	5
2	2	13	13	11	10	9
3	9	15	19	11	20	8
4	3	11	15	12	10	16
5	4	19	20	10	14	13
6	7	12	17	11	no	12
7	8	20	14	10	14	12
8	5	16	12	17	30	11
9	1	14	18	7	12	14
10	6	18	11	7	1	17

Answers

Unit 5 Subtraction: Subtracting multiples of 10

	Set A	Set B	Set C	Set D	Set E	Set F
1	7	26	114	14	17	15
2	8	27	157	7	42	$17
3	9	54	114	46	64	47
4	2	21	108	57	36	44
5	4	58	135	4	51	39
6	5	14	168	27	64	27
7	8	75	141	58	44	63
8	7	36	142	21	42	14
9	1	73	171	34	67	33
10	3	29	148	32	61	69

Unit 6 Subtraction: Subtracting 1–9

	Set A	Set B	Set C	Set D	Set E	Set F
1	6	3	25	36	38	95
2	2	15	68	24	62	89
3	7	25	27	48	51	87
4	8	63	75	34	26	79
5	5	56	17	87	19	72
6	12	86	77	73	79	63
7	14	35	86	38	9	58
8	21	47	21	79	58	49
9	34	62	42	68	78	43
10	44	65	74	23	68	36

Unit 7 Doubling and near doubling

	Set A	Set B	Set C	Set D	Set E	Set F
1	14	16	48	50	13	$68
2	6	20	66	90	15	32
3	22	14	88	30	23	26
4	18	16	46	70	29	36
5	10	24	42	32	43	68
6	16	24	62	54	65	54
7	8	18	44	72	52	78
8	12	20	82	96	86	104
9	20	22	68	58	63	$52
10	24	22	86	74	41	40

Unit 8 Halving and near halving

	Set A	Set B	Set C	Set D	Set E	Set F
1	6	15	121	$1\frac{1}{2}$	23	84
2	11	40	144	$3\frac{1}{2}$	61	33
3	23	20	221	$5\frac{1}{2}$	210	15
4	41	35	320	$7\frac{1}{2}$	124	104
5	33	50	444	$10\frac{1}{2}$	88	7
6	24	45	133	$14\frac{1}{2}$	666	42
7	31	30	240	$15\frac{1}{2}$	5	$5.50
8	44	25	344	$18\frac{1}{2}$	11	8
9	22	55	404	$21\frac{1}{2}$	41	60
10	34	60	304	$24\frac{1}{2}$	101	1000

Answers

Unit 9 Multiplication: The 6 times and 12 times tables

	Set A	Set B	Set C	Set D	Set E	Set F
1	42	48	26	6, 12	24	42
2	12	96	39	6	48	48
3	66	84	67	neither	12 is twice 6	120
4	30	144	40	6, 12	18	36
5	54	36	80	6	120	30
6	72	72	46	neither	no	66
7	18	60	18	6, 12	48	54
8	48	108	14	6, 12	88	60
9	24	24	115	6	34	42
10	36	120	39	6, 12	136	24

Unit 10 Multiplication: The 7 times and 8 times tables

	Set A	Set B	Set C	Set D	Set E	Set F
1	35	88	23	7	49	40
2	49	40	42	neither	32	77
3	77	24	82	7	84	24
4	28	96	51	8	48	64
5	56	48	82	7 and 8	64	49
6	84	72	60	neither	9	72
7	42	80	43	7	12	42
8	70	56	40	8	56	40
9	21	32	78	neither	21, 24, 28	70
10	63	64	81	7	77	63

Unit 11 Multiplication: The 9 times and 11 times tables

	Set A	Set B	Set C	Set D	Set E	Set F
1	63	55	46	9	33	90
2	36	88	60	11	63	$55
3	99	22	75	neither	88	27
4	45	121	101	11	36	77
5	72	77	85	9, 11	72	88
6	108	66	71	9	22	90
7	27	44	35	11	10	66
8	54	110	101	neither	12	27
9	90	99	100	9	50	110
10	81	132	29	neither	34	99

Unit 12 Division: Dividing by 6 and 12

	Set A	Set B	Set C	Set D	Set E	Set F
1	6, 12	4	84	3	5	12
2	6	6	132	3	9	11
3	6, 12	11	24	2	10	5
4	6	8	108	1	3	4
5	no	2	0	3	11	8
6	6, 12	5	60	1	8	7
7	no	12	120	4	0	9
8	6	7	96	1	4	5, 7
9	6	1	48	2	5	11
10	no	9	72	5	4	3

Answers

Holy Mighty Mentals, it's Super Brain!

Unit 13 Division: Dividing by 7 and 8

	Set A	Set B	Set C	Set D	Set E	Set F
1	12	10	2	1	6	3
2	7	3	6	7	11	7
3	10	9	0	3	9	4
4	3	4	3	5	12	11
5	4	12	2	0	4	5
6	11	6	5	6	5	6
7	8	5	4	2	1	9
8	2	11	0	0	2	8
9	9	8	4	4	1	12
10	6	7	6	6	12	12

Unit 14 Division: Dividing by 9 and 11

	Set A	Set B	Set C	Set D	Set E	Set F
1	11	5	5	4	5	7
2	8	9	8	10	8	12
3	6	6	1	3	3	8
4	4	12	6	8	12	6
5	12	3	0	0	4	6
6	2	8	5	1	8	9
7	9	4	4	5	12	2
8	7	2	7	0	6	4
9	5	11	2	7	5	5
10	10	10	0	2	4	9

Unit 15 Mixed multiplication and division

	Set A	Set B	Set C	Set D	Set E	Set F
1	28	6	40	2r1	14	28
2	88	7	120	4r2	37	7
3	45	9	36	2r1	11	24
4	72	9	18	5r3	24	66
5	30	8	40	6r3	2	27
6	2	7	27	10r1	2	$90
7	63	6	100	4r5	40	12
8	20	11	96	11r1	3	42
9	48	7	60	7r2	72	8
10	36	9	108	7r7	24	84

Unit 16 Odd and even numbers

	Set A	Set B	Set C	Set D	Set E	Set F
1	even	5	10	even	even	10
2	odd	11	24	odd	even	9
3	odd	23	238	even	odd	22
4	even	45	548	odd	odd	21
5	even	99	788	even	even	40
6	odd	161	900	even	even	39
7	even	275	1000	even	even	66
8	odd	483	1028	even	odd	65
9	even	1091	2388	even	odd	100
10	odd	8001	3880	odd	even	99

Answers

Unit 17 Place value: Digit recognition

Set A	Set B	Set C	Set D	Set E	Set F
1 H	1 2732	1 7040	1 130	1 9	1 100
2 U	2 3179	2 603	2 27	2 99	2 10
3 TH	3 4050	3 8307	3 7300	3 999	3 100
4 T	4 1009	4 9710	4 23 000	4 9999	4 100
5 H	5 8201	5 5382	5 890	5 10	5 1000
6 TH	6 6073	6 9030	6 9300	6 100	6 10
7 T	7 2080	7 10 006	7 18 000	7 1000	7 100
8 T	8 3108	8 20 700	8 53	8 10	8 1000
9 TH	9 6249	9 7034	9 410	9 100	9 10
10 T	10 8941	10 80 008	10 10 000	10 10	10 100

Unit 18 Rounding off: Nearest 10, 100 and 1000

Set A	Set B	Set C	Set D	Set E	Set F
1 75	1 1500	1 100	1 1000	1 1	1 1499
2 125	2 2500	2 100	2 1000	2 450	2 4499
3 365	3 3500	3 300	3 1000	3 750	3 8499
4 450	4 4500	4 300	4 2000	4 350	4 3499
5 950	5 5500	5 500	5 2000	5 650	5 9499
6 2750	6 6500	6 500	6 4000	6 150	6 5499
7 3450	7 7500	7 700	7 5000	7 250	7 7499
8 7885	8 8500	8 800	8 5000	8 950	8 6499
9 8050	9 9500	9 900	9 7000	9 1050	9 2499
10 10 050	10 10 500	10 900	10 9000	10 1450	10 10 499

Unit 19 Sequences: Finding terms and rules

Set A	Set B	Set C	Set D	Set E	Set F
1 16	1 56	1 +6	1 –5	1 I	1 16
2 20	2 88	2 +4	2 –10	2 M	2 10
3 14	3 28	3 +10	3 –8	3 K	3 12
4 8	4 45	4 +11	4 –4	4 U	4 Apr
5 27	5 42	5 +5	5 –12	5 Y	5 autumn
6 22	6 100, 90	6 +8	6 –7	6 F	6 evif
7 12, 24	7 36, 24	7 +7	7 –9	7 O	7 pentagon
8 21, 28	8 33, 30	8 +9	8 –3	8 X	8 Sat
9 10, 20	9 108, 84	9 +12	9 –6	9 W	9 10 000
10 12, 24, 48	10 90, 81	10 +20	10 –11	10 P	10 S (six)

Unit 20 Number strings

Set A	Set B	Set C	Set D	Set E	Set F
1 7	1 46	1 5	1 17	1 1	1 44
2 14	2 28	2 12	2 30	2 24	2 17
3 19	3 47	3 15	3 25	3 5	3 16
4 13	4 36	4 8	4 14	4 40	4 10
5 21	5 63	5 19	5 11	5 1	5 17
6 26	6 59	6 3	6 60	6 10	6 19
7 14	7 51	7 3	7 55	7 11	7 44
8 11	8 5	8 0	8 57	8 9	8 $55
9 10	9 40	9 2	9 12	9 2	9 3
10 19	10 12	10 3	10 7	10 30	10 16

Answers

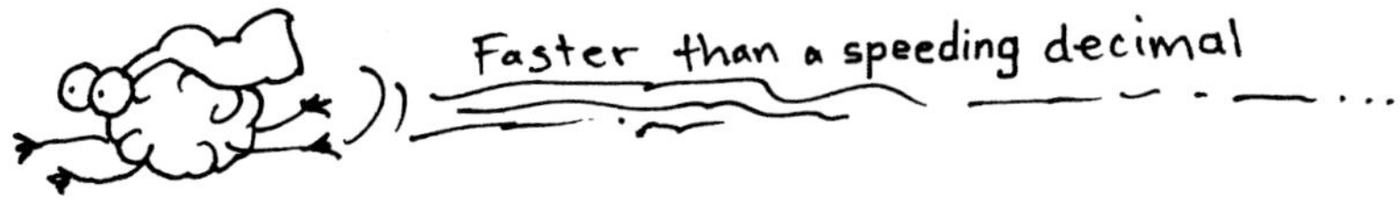

Unit 21 Fractions and decimals

	Set A	Set B	Set C	Set D	Set E	Set F
1	7	$\frac{1}{2}$	2	$\frac{1}{10}$	1	4
2	4	$\frac{1}{4}$	4	$\frac{4}{10}$	6	10
3	5	$\frac{1}{2}$	10	0.7	8	6
4	28	$\frac{1}{10}$	4	0.9	4	12
5	20	$\frac{1}{4}$	8	0.5	10	30
6	70	$\frac{1}{4}$	20	$\frac{8}{10}$	7	6
7	20	$\frac{1}{10}$	8	$\frac{2}{10}$	2	9
8	10	$\frac{1}{4}$	40	$\frac{1}{2}$	5	15
9	48	$\frac{1}{2}$	60	$\frac{3}{10}$	3	8
10	$\frac{1}{4}$	$\frac{1}{10}$	100	0.6	9	27

Unit 22 Length: Units and conversions

	Set A	Set B	Set C	Set D	Set E	Set F
1	700	4000	20	2	metres	23
2	1000	7000	50	4	centimetres	7
3	300	10 000	100	5	centimetres	4
4	50	500	5	1	millimetres	8000
5	150	3500	75	2	kilometres	15
6	7	3	6	20	centimetres	5
7	5	6	9	40	centimetres	10
8	10	10	4	100	kilometres	false
9	$2\frac{1}{2}$	$\frac{1}{2}$	$\frac{1}{2}$	10	millimetres	false
10	$9\frac{1}{2}$	$2\frac{1}{2}$	$1\frac{1}{2}$	15	metres	2000

Unit 23 Perimeter: Concept and missing sides

	Set A	Set B	Set C	Set D (in any order)	Set E	Set F
1	16 cm	3 mm	20 mm	L = 21 cm; W = 1 cm	15 cm	2
2	44 cm	4 mm	32 mm	L = 20 cm; W = 2 cm	20 cm	100 cm
3	100 cm	6 mm	28 mm	L = 19 cm; W = 3 cm	25 cm	30 m
4	28 mm	2 cm	26 cm	L = 18 cm; W = 4 cm	30 cm	metres
5	48 mm	7 cm	62 cm	L = 17 cm; W = 5 cm	35 cm	true
6	20 m	11 cm	40 cm	L = 16 cm; W = 6 cm	40 cm	true
7	32 m	5 m	10 m	L = 15 cm; W = 7 cm	45 cm	false
8	36 km	12 m	40 m	L = 14 cm; W = 8 cm	50 cm	metres
9	60 km	100 km	6 km	L = 13 cm; W = 9 cm	60 cm	true
10	80 km	212 km	18 km	L = 12 cm; W = 10 cm	100 cm	false

Unit 24 Area: Concept and application

	Set A	Set B	Set C	Set D	Set E	Set F
1	6 sq cm	144 sq m	5 cm	sq m	sq m	Western Australia
2	50 sq cm	16 sq m	9 cm	sq m	sq m	Tasmania
3	32 sq cm	49 sq m	2 cm	sq cm	sq m	false
4	72 sq cm	81 sq m	2 cm	sq cm	sq cm	true
5	35 sq cm	121 sq m	4 cm	sq cm	sq cm	true
6	88 sq cm	9 sq m	6 m	sq cm	sq cm	true
7	72 sq cm	64 sq m	6 m	sq cm	sq m	false
8	80 sq cm	100 sq m	4 m	sq m	sq cm	area
9	60 sq cm	36 sq m	11 m	sq m	sq m	false
10	1000 sq cm	25 sq m	10 m	sq cm	sq m	true

Answers

Unit 25 Mass: Units and conversions

	Set A	Set B	Set C	Set D	Set E	Set F
1	5 kg	3000 g	100 g	200 g	grams	3000
2	3 kg	4000 g	900 g	250 g	grams	2
3	9 kg	8000 g	500 g	300 g	kilograms	10
4	11 kg	12 000 g	950 g	6 kg	grams	5
5	20 kg	30 000 g	750 g	15 g	kilograms	false
6	$\frac{1}{2}$ kg	500 g	150 g	1 kg	kilograms	scales
7	$1\frac{1}{2}$ kg	2500 g	25 g	200 g	grams	gram
8	$4\frac{1}{2}$ kg	4500 g	980 g	2 kg	grams	3
9	$7\frac{1}{2}$ kg	6500 g	650 g	5 kg	kilograms	lighter
10	$10\frac{1}{2}$ kg	8500 g	115 g	750 g	grams	true

Unit 26 Capacity/volume: Units and conversions

	Set A	Set B	Set C	Set D	Set E	Set F
1	2 L	3000 mL	200 mL	mL	1 kg	36 kg
2	4 L	6000 mL	750 mL	mL	500 g	750
3	7 L	8000 mL	8 L	L	1500 g	4
4	1 L	5000 mL	50 mL	L	4 kg	8200
5	10 L	9000 mL	2 L	L	4500 g	500
6	$1\frac{1}{2}$ L	2500 mL	40 L	L	8 kg	2000
7	$3\frac{1}{2}$ L	4500 mL	2000 L	mL	10 kg	250
8	$8\frac{1}{2}$ L	9500 mL	300 mL	mL	100 kg	400
9	$5\frac{1}{2}$ L	7500 mL	$1\frac{1}{2}$ L	L	1000 kg	3
10	$10\frac{1}{2}$ L	6500 mL	50 L	mL	10 000 kg	20

Unit 27 Time: Digital and analogue conversions

	Set A	Set B	Set C	Set D	Set E	Set F
1	25 minutes past 3	8:23	5 minutes to 7	5:50	20 minutes	25
2	9 minutes past 10	11:17	20 minutes to 9	3:36	10 minutes	5:05
3	22 minutes past 5	6:07	12 minutes to 12	11:52	20 minutes	2:40
4	17 minutes past 2	4:12	25 minutes to 5	4:40	25 minutes	8:05
5	14 minutes past 8	10:28	1 minute to 3	2:58	13 minutes	11:45
6	27 minutes past 12	7:05	29 minutes to 6	10:53	25 minutes	10 minutes past 3
7	4 minutes past 6	2:19	11 minutes to 1	1:59	13 minutes	12:10
8	21 minutes past 1	12:22	14 minutes to 2	6:46	14 minutes	7:58
9	19 minutes past 4	3:14	3 minutes to 4	12:32	9 minutes	1:35
10	1 minute past 7	8:29	27 minutes to 9	7:31	14 minutes	5:59

Unit 28 Time: Unit equivalences

	Set A	Set B	Set C	Set D	Set E	Set F
1	120	2	30 days	autumn	December	130
2	4	35	30 days	winter	October	February
3	180	700	31 days	summer	March or April	120
4	90	11	31 days	spring	May	90
5	half	2	31 days	winter	September	September, October, November
6	quarter	20	31 days	autumn	December	February
7	48	104	31 days	summer	February	$520
8	10	520	31 days	spring	January	16
9	2400	2	28 or 29 days	winter	December	16
10	36	120	30 days	autumn	January	3

Answers

Unit 29 Money: Economical ways of making amounts

	Set A	Set B	Set C	Set D	Set E	Set F
1	2	2	2	2	3	2
2	2	3	3	2	4	$12.50
3	3	4	1	3	5	2, 2
4	1	2	2	4	5	$20.05
5	2	3	3	5	4	$20 + 5 cents
6	3	3	2	3	7	85 cents
7	4	5	3	5	3	$3.85
8	3	2	4	4	3	$185
9	4	3	3	6	4	$50 + $20 + $10 + $5
10	1	6	4	8	6	$100 + $100 + $20 + $20 + $5 + 50 cents + 20 cents + 5 cents

Unit 30 2D and 3D: Properties and directions

	Set A	Set B	Set C	Set D	Set E	Set F
1	6	6	12	8	east	true
2	12	6	12	8	west	false
3	12	3	2	0	south	true
4	20	5	9	6	south	square and rectangle
5	25	8	18	12	west	triangle and rectangle
6	12	1	0	0	east	false
7	21	5	8	5	north	hexagons
8	32	7	12	7	north	octagons
9	18	2	1	0	south	cylinder
10	1000	9	16	9	south	rectangular prism

Unit 31 Chance: Concepts of likelihood

	Set A	Set B	Set C	Set D	Set E	Set F
1	5	possible	a card less than 7	$\frac{1}{10}$	true	4
2	5	50/50	50/50	$\frac{1}{10}$	true	$\frac{4}{10}$
3	10	impossible	50/50	$\frac{5}{10}$	true	6
4	8	possible	50/50	$\frac{5}{10}$	false	$\frac{6}{10}$
5	4	certain	a card greater than 4	$\frac{2}{10}$	false	2
6	6	50/50	a card greater than 6	$\frac{3}{10}$	false	$\frac{2}{10}$
7	4	50/50	an even card	$\frac{3}{10}$	true	3
8	1	certain	a card greater than 1	$\frac{2}{10}$	true	$\frac{3}{10}$
9	2	impossible	an odd red card	$\frac{2}{10}$	false	1
10	2	50/50	50/50	$\frac{9}{10}$	false	$\frac{1}{10}$

Unit 32 Revision: All sorts

	Set A	Set B	Set C	Set D	Set E	Set F
1	14	300	1	$20 + $10 + $5 + $1	1700	23
2	9	48	28	sphere	13	12
3	46	2	6	east	$\frac{7}{10}$	72
4	6	20	10	50/50	3000	3
5	9	4	16	34	35	108
6	23	45	3	69	5000	1 whole number
7	4	$\frac{1}{8}$	40 sq cm	72	Friday	3, 2
8	130	64	12	7	4	30, winter
9	39	36	autumn	23	east	24
10	6	64	28 days	200	$\frac{1}{6}$	$\frac{5}{15}$

Set C

1 5 x 2 x 4 = ☐
2 6 x 2 x 10 = ☐
3 4 x 3 x 3 = ☐
4 3 x 3 x 2 = ☐
5 10 x 4 x 1 = ☐
6 3 x 3 x 3 = ☐
7 10 x 5 x 2 = ☐
8 6 x 2 x 8 = ☐
9 4 x 3 x 5 = ☐
10 12 x 1 x 9 = ☐

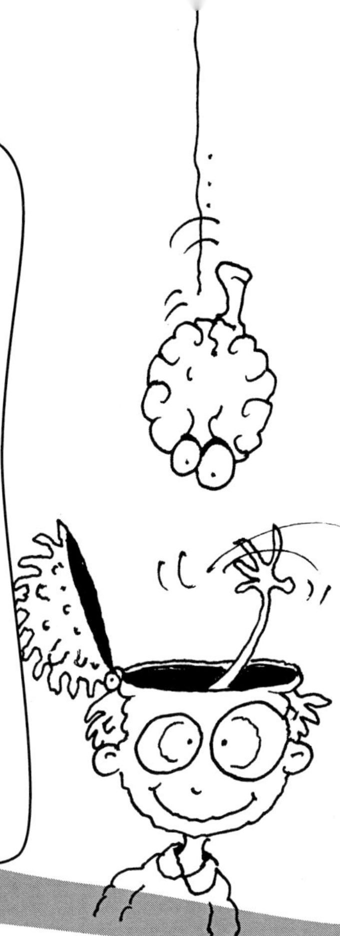

Set D

1 13 ÷ 6 = ☐
2 46 ÷ 11 = ☐
3 19 ÷ 9 = ☐
4 23 ÷ 4 = ☐
5 33 ÷ 5 = ☐
6 21 ÷ 2 = ☐
7 37 ÷ 8 = ☐
8 100 ÷ 9 = ☐
9 51 ÷ 7 = ☐
10 77 ÷ 10 = ☐

Set F

1 My 7 pet mice have ____ legs altogether.
2 A 50 cent coin has 12 sides. There are 84 sides on ____ 50 cent coins.
3 There are ____ sides on 6 squares.
4 11 goals in Australian football equal ____ points.

5 9 tricycles have ____ wheels.
6 A wallet containing nine $10 notes holds _____ in total.
7 An hour is made up of ____ 5 minute periods of time.
8 7 packets of muffins, each containing 6 muffins, equals ____ muffins.
9 Buying $40 worth of $5 books will get you ____ books.
10 A 7-year drought lasts for ____ months.

Unit 16 Odd and even numbers

Odd numbers end in 1, 3, 5, 7 or 9. Even numbers end in 0, 2, 4, 6 or 8. The third odd number comes before the third even number: 5 before 6. The 100th odd number comes before the 100th even number: 199 before 200, and so on.

Set A

Are the following numbers odd or even?

1 **4** ______
2 **7** ______
3 **11** ______
4 **28** ______
5 **36** ______
6 **45** ______
7 **62** ______
8 **73** ______
9 **80** ______
10 **99** ______

Set B

What is the next odd number after:

1 **3?** ☐
2 **10?** ☐
3 **21?** ☐
4 **44?** ☐
5 **97?** ☐
6 **160?** ☐
7 **274?** ☐
8 **481?** ☐
9 **1090?** ☐
10 **7999?** ☐

Set E

Will the answers to the following questions be odd or even?

1 4 x 12 ______
2 3 x 4 ______
3 7 x 5 ______
4 11 x 9 ______
5 12 x 6 ______
6 12 ÷ 6 ______
7 24 ÷ 3 ______
8 81 ÷ 9 ______
9 121 ÷ 11 ______
10 36 ÷ 6 ______

Set C

What is the next even number after:

1	**9?**	
2	**22?**	
3	**237?**	
4	**546?**	
5	**787?**	
6	**898?**	
7	**999?**	
8	**1027?**	
9	**2386?**	
10	**3879?**	

Set D

Will the answers to these questions be odd or even?

1 **4 + 8** ________

2 **12 + 7** ________

3 **5 + 9** ________

4 **23 + 12** ________

5 **37 + 25** ________

6 **20 – 8** ________

7 **71 – 9** ________

8 **100 – 22** ________

9 **1000 – 100** ________

10 **2000 – 10 – 1** ________

Set F

1 What is the 5th even number? ______

2 What is the 5th odd number? ______

3 What is the 11th even number? ______

4 What is the 11th odd number? ______

5 What is the 20th even number? ______

6 What is the 20th odd number? ______

7 What is the 33rd even number? ______

8 What is the 33rd odd number? ______

9 What is the 50th even number? ______

10 What is the 50th odd number? ______

Number and Algebra

Unit 17
Place value: Digit recognition

We know that our counting system is based on the number 10, because we have 10 fingers. So, each time we have 10 of a basic unit, we need to introduce a new value. 10 ones or units make 10; 10 tens make 100; 10 hundreds make 1000, etc.

Set A

What is the 3 worth in the following numbers? Write 'U' for units, 'T' for tens, 'H' for hundreds and 'TH' for thousands.

1 4326 ☐
2 5653 ☐
3 3870 ☐
4 2931 ☐
5 7329 ☐
6 3487 ☐
7 2138 ☐
8 8031 ☐
9 3492 ☐
10 8030 ☐

Set B

Write down the values of the following numbers.

1 **2000 + 700 + 30 + 2** ☐
2 **3000 + 100 + 70 + 9** ☐
3 **4000 + 50** ☐
4 **1000 + 9** ☐
5 **8000 + 200 + 1** ☐
6 **6000 + 70 + 3** ☐
7 **80 + 2000** ☐
8 **100 + 3000 + 8** ☐
9 **40 + 200 + 9 + 6000** ☐
10 **8000 + 40 + 900 + 1** ☐

Set E

1 What is 1 less than 10? ______
2 What is 1 less than 100? ______
3 What is 1 less than 1000? ______
4 What is 1 less than 10 000? ______
5 How many tens make a hundred? ______
6 How many tens make a thousand? ______
7 How many tens make ten thousand? ______
8 How many hundreds make a thousand? ______
9 How many hundreds make ten thousand? ______
10 How many thousands make ten thousand? ______

Set C

What do the following numbers combine to make?

1 4 tens + 7 thousands ________

2 3 ones + 6 hundreds ________

3 8 thousands + 3 hundreds + 7 units ________

4 1 ten + 7 hundreds + 9 thousands ________

5 5 thousands + 3 hundreds + 8 tens + 2 ones ________

6 3 tens + 9 thousands ________

7 6 ones + 10 thousands ________

8 7 hundreds + 20 thousands ________

9 3 tens + 7 thousands + 4 units ________

10 80 thousands + 8 ones ________

Set D

What are the values of the following numbers?

1 13 tens ________

2 27 units ________

3 73 hundreds ________

4 23 thousands ________

5 89 tens ________

6 93 hundreds ________

7 18 thousands ________

8 53 ones ________

9 41 tens ________

10 100 hundreds ________

Set F

1 I have $100 in $1 coins. I have ________ $1 coins.

2 I have $1 in 10 cent coins. I have ________ 10 cent coins.

3 I have $10 in 10 cent coins. I have ________ 10 cent coins.

4 There are ________ centimetres in 1 metre.

5 The snail crawled 10 metres. It crawled ________ centimetres.

6 How many decades are in a century? ________

7 There are ________ decades in 1000 years.

8 I trained by running ten 100 metre sprints. I ran ________ metres altogether.

9 How many $10 notes equal $100? ________

10 There are ________ 100 metre runs in the 10 000 metre race.

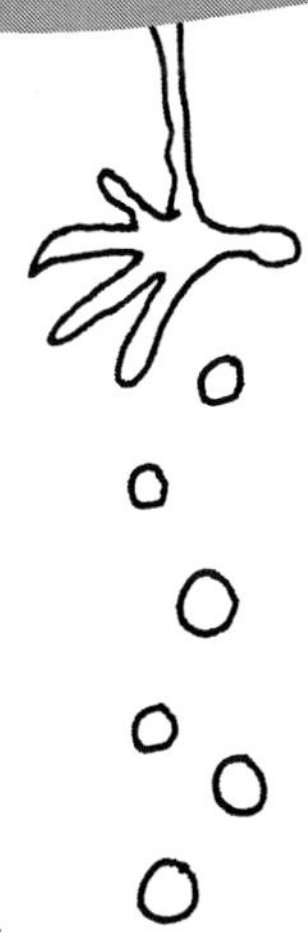

Number and Algebra

Unit 18
Rounding off: Nearest 10, 100 and 1000

When numbers are rounded off, they are taken to the nearest 10, 100 or 1000, depending on the question. If a number is exactly halfway, it is taken up. So, 25 to the nearest 10 equals 30; 650 to the nearest 100 is 700; 5500 to the nearest 1000 is 6000.

Set A

What number is halfway between:

1 **70 and 80?** ________
2 **120 and 130?** ________
3 **360 and 370?** ________
4 **400 and 500?** ________
5 **900 and 1000?** ________
6 **2700 and 2800?** ________
7 **3400 and 3500?** ________
8 **7880 and 7890?** ________
9 **8000 and 8100?** ________
10 **10 000 and 10 100?** ________

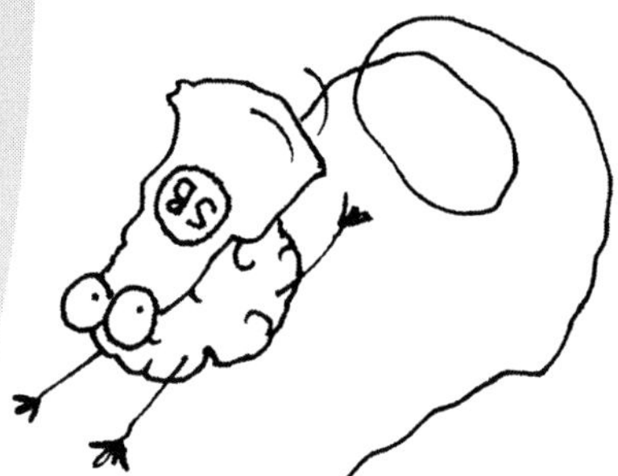

Set D

Round the following numbers to the nearest 1000.

1	**624**	
2	**1199**	
3	**1455**	
4	**1800**	
5	**2499**	
6	**3902**	
7	**5050**	
8	**5450**	
9	**7219**	
10	**8500**	

Set E

What is the smallest number that, when rounded to the nearest 100, will equal:

1 100? ________
2 500? ________
3 800? ________
4 400? ________
5 700? ________
6 200? ________
7 300? ________
8 1000? ________
9 1100? ________
10 1500? ________

Set B

What number is halfway between:

1 **1000 and 2000?** ________

2 **2000 and 3000?** ________

3 **3000 and 4000?** ________

4 **4000 and 5000?** ________

5 **5000 and 6000?** ________

6 **6000 and 7000?** ________

7 **7000 and 8000?** ________

8 **8000 and 9000?** ________

9 **9000 and 10 000?** ________

10 **10 000 and 11 000?** ________

Set C

Round the following numbers to the nearest 100.

1 **97** ☐

2 **149** ☐

3 **251** ☐

4 **345** ☐

5 **491** ☐

6 **509** ☐

7 **682** ☐

8 **837** ☐

9 **850** ☐

10 **944** ☐

Set F

What is the biggest number that, when rounded to the nearest 1000, will equal:

1 1000? ________

2 4000? ________

3 8000? ________

4 3000? ________

5 9000? ________

6 5000? ________

7 7000? ________

8 6000? ________

9 2000? ________

10 10 000? ________

Number and Algebra

Unit 19
Sequences: Finding terms and rules

A sequence is a pattern including numbers, letters or symbols. Each member of a sequence is called a term, and each sequence's pattern is called a rule. The sequence 3, 6, 9 . . . has 9 as its 3rd term, and +3 or the 3 times table as its rule.

Set A

Find the missing terms in the following sequences.

1 4, 8, 12, ___
2 5, 10, 15, ___
3 7, ___, 21, 28
4 ___, 16, 24, 32
5 9, 18, ___, 36
6 11, ___, 33, 44
7 6, ___, 18, ___, 30
8 7, 14, ___, ___, 35
9 ___, ___, 30, 40
10 ___, ___, 36, ___, 60

Set B

What are the missing terms in the following sequences?

1 80, 72, 64, ___
2 121, 110, 99, ___
3 32, ___, 24, 20
4 55, 50, ___, 40
5 49, ___, 35, 28
6 110, ___, ___, 80
7 48, 42, ___, 30, ___
8 ___, ___, 27, 24, 21
9 ___, 96, ___, 72
10 99, ___, ___, 72

Set E

Find the missing terms in the following sequences.

1 A, C, E, G, ___
2 A, D, G, J, ___
3 A, B, D, G, ___
4 A, E, I, O, ___
5 A, G, M, S, ___
6 J, I, H, G, ___
7 W, U, S, Q, ___
8 B, A, Z, Y, ___
9 E, C, A, Y, ___
10 Z, Y, W, T, ___

Set C

What are the rules of the following sequences?

1 **7, 13, 19, 25** ____

2 **23, 27, 31, 35** ____

3 **78, 88, 98, 108** ____

4 **9, 20, 31, 42** ____

5 **37, 42, 47, 52** ____

6 **12, 20, 28, 36** ____

7 **32, 39, 46, 53** ____

8 **2, 11, 20, 29** ____

9 **17, 29, 41, 53** ____

10 **15, 35, 55, 75** ____

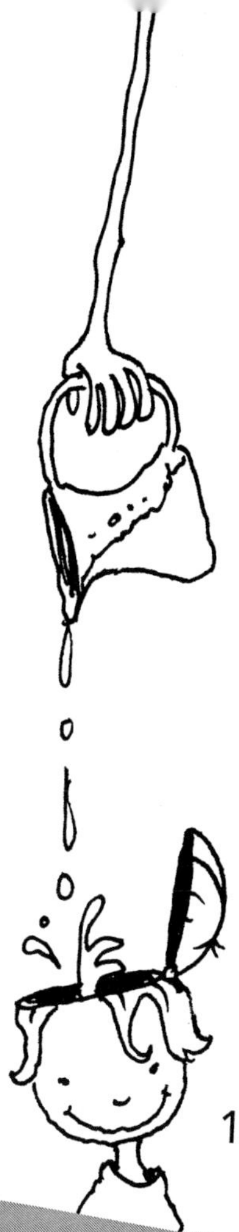

Set D

Find the rules of the following sequences.

1 **64, 59, 54, 49** ____

2 **87, 77, 67, 57** ____

3 **86, 78, 70, 62** ____

4 **21, 17, 13, 9** ____

5 **100, 88, 76, 64** ____

6 **60, 53, 46, 39** ____

7 **93, 84, 75, 66** ____

8 **16, 13, 10, 7** ____

9 **50, 44, 48, 32** ____

10 **83, 72, 61, 50** ____

Set F

Find the missing terms in the following sequences.

1 1, 2, 4, 8 ___

2 80, 40, 20, ___

3 2, 3, 5, 8, ___

4 Jan, Feb, Mar, _____

5 spring, summer, __________, winter

6 eno, owt, eerht, ruof, _____

7 triangle, square, ________________, hexagon

8 Sun, Tue, Thu, _____

9 1, 10, 100, 1000, _________

10 O, T, T, F, F, ______

Unit 20 Number strings

Number strings are questions made up of many smaller pieces. Work left to right to solve these challenging problems.

Set A

1 6 + 5 – 4 =
2 10 + 8 – 4 =
3 20 + 4 – 5 =
4 11 + 9 – 7 =
5 20 + 8 – 7 =
6 25 – 5 + 6 =
7 13 – 7 + 8 =
8 16 – 7 + 2 =
9 9 – 7 + 8 =
10 14 – 4 + 9 =

Set B

1 7 x 6 + 4 =
2 5 x 5 + 3 =
3 10 x 4 + 7 =
4 9 x 3 + 9 =
5 12 x 5 + 3 =
6 11 x 6 – 7 =
7 8 x 7 – 5 =
8 2 x 8 – 11 =
9 7 x 7 – 9 =
10 3 x 8 – 12 =

Set E

1 20 ÷ 5 ÷ 4 = ___
2 3 x 2 x 4 = ___
3 100 ÷ 10 ÷ 2 = ___
4 4 x 5 x 2 = ___
5 60 ÷ 12 ÷ 5 = ___
6 2 x 2 x 3 – 2 = ___
7 30 ÷ 5 + 6 – 1 = ___
8 15 ÷ 5 x 3 = ___
9 8 x 2 ÷ 4 – 2 = ___
10 7 x 7 – 10 – 9 = ___

Set C

1 8 ÷ 4 + 3 = ☐
2 12 ÷ 3 + 8 = ☐
3 20 ÷ 4 + 10 = ☐
4 24 ÷ 6 + 4 = ☐
5 100 ÷ 10 + 9 = ☐
6 22 ÷ 2 – 8 = ☐
7 36 ÷ 6 – 3 = ☐
8 42 ÷ 6 – 7 = ☐
9 18 ÷ 3 – 4 = ☐
10 90 ÷ 9 – 7 = ☐

Set D

1 5 + 5 + 5 + 2 = ☐
2 10 + 10 + 4 + 6 = ☐
3 6 + 6 + 3 + 10 = ☐
4 10 – 5 + 7 + 2 = ☐
5 12 + 12 – 11 – 2 = ☐
6 100 – 10 – 10 – 20 = ☐
7 100 – 20 – 20 – 5 = ☐
8 100 – 50 + 7 = ☐
9 40 – 20 – 10 + 2 = ☐
10 60 – 30 – 20 – 10 + 7 = ☐

Set F

1 Multiply 5 lollies by 10 and eat 6. ____ lollies left.
2 Double 12 cats and then give away 7. ____ cats left.
3 Halve 24 dogs and buy 4 more. ____ dogs left.
4 Divide 20 fish by 4 and double the result. ____ fish left.
5 Kick 5 goals, then 5 more and then 6 more and lose by 1 goal. The winners kicked ____ goals.
6 Score a 6, a 4, a 3, another 6 and then get out. You scored ____ runs.
7 Catch 11 flies that double in number and then double again. There are now ____ flies.
8 Bring $100, spend $20 and then spend $15 and then spend $10. ____ left in change.
9 Catch 12 rabbits, let 6 go and give half of the rest away. ____ rabbits left.
10 Start with 20 marbles, lose 8, then lose 4 and then double what is left. You now have ____ marbles.

Number and Algebra

Unit 21: Fractions and decimals

A fraction is like a division problem: $\frac{1}{4}$ means split or divide something up into 4 equal pieces and take one of those equal pieces.

A decimal is a fraction that deals with splitting things up into 10, 100 or 1000 equal pieces and is written differently, using a point.

Set A

1 What is $\frac{1}{2}$ of 14? ___
2 What is $\frac{1}{4}$ of 16? ___
3 What is $\frac{1}{10}$ of 50? ___
4 14 is a half of ___
5 5 is a quarter of ___
6 7 is a tenth of ___
7 2 is a tenth of ___
8 What is $\frac{1}{4}$ of 40? ___
9 12 is $\frac{1}{4}$ of ___
10 $\frac{1}{2}$ of a half is ___

Set D

1 What fraction equals 0.1? ___
2 0.4 equals the fraction ___
3 $\frac{7}{10}$ as a decimal is ___
4 $\frac{9}{10}$ as a decimal is ___
5 $\frac{1}{2}$ as a decimal is ___
6 0.8 equals the fraction ___
7 0.2 equals the fraction ___
8 0.5 equals the fraction ___
9 0.3 equals the fraction ___
10 6/10 equals the decimal ___

Set E

1 0.1 of 10 = ___
2 0.1 of 60 = ___
3 0.1 of 80 = ___
4 0.1 of 40 = ___
5 0.1 of 100 = ___
6 0.1 of 70 = ___
7 0.1 of 20 = ___
8 0.1 of 50 = ___
9 0.1 of 30 = ___
10 0.1 of 90 = ___

Set B

1 What fraction of 20 is 10? ___

2 What fraction of 40 is 10? ___

3 What fraction of 100 is 50? ___

4 What fraction of 100 is 10? ___

5 What fraction of 16 is 4? ___

6 What fraction of 80 is 20? ___

7 What fraction of 90 is 9? ___

8 What fraction of 100 is 25? ___

9 What fraction of 1000 is 500? ___

10 What fraction of 1000 is 100? ___

Set C

1 How many halves make a whole number? ____

2 How many quarters make a whole? ____

3 How many tenths make a whole number? ____

4 How many halves make 2 whole numbers? ____

5 How many quarters make 2 whole numbers? ____

6 How many tenths make 2 whole numbers? ____

7 How many halves make 4 whole numbers? ____

8 How many quarters make 10 whole numbers? ____

9 How many tenths make 6 whole numbers? ____

10 How many halves make 50 whole numbers? ____

Set F

1 I scored half of our team's 8 goals. I scored ____ goals.

2 A quarter of the 40 minute netball game was played in rain. It rained for ____ minutes.

3 Our team won one-tenth of the 60 swimming races. We won ____ races.

4 I slept a half of the day away. I slept for ____ hours.

5 A quarter of my 40 answers were even. ____ answers were odd.

6 A half of my 12 pairs of shoes are black. ____ pairs are *not* black.

7 One-tenth of our 90 runs were in singles. ____ runs were in singles.

8 I read for a quarter of an hour. I read for ____ minutes.

9 I have read one-tenth of the 80-page book. I have read ____ pages.

10 It was hot for only one-tenth of the days in November. It was *not* hot for ____ days.

Measurement and Geometry

Unit 22 Length: Units and conversions

Remember that, in measurement, 10 mm = 1 cm; 100 cm = 1 m and 1000 m = 1 km.

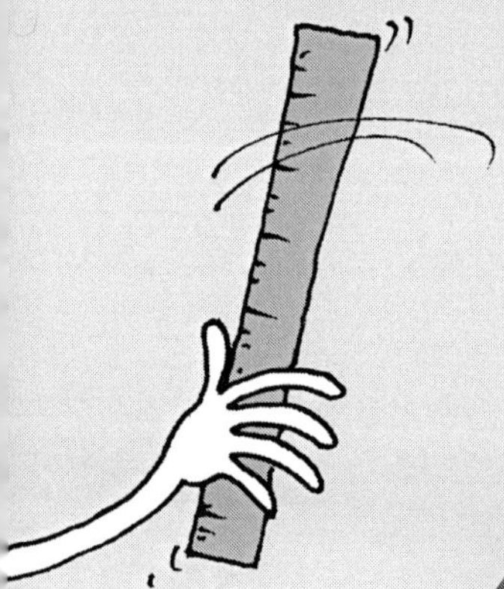

Set A

1 7 m = ______ cm

2 10 m = ______ cm

3 3 m = ______ cm

4 $\frac{1}{2}$ m = ______ cm

5 $1\frac{1}{2}$ m = ______ cm

6 700 cm = ______ m

7 500 cm = ______ m

8 1000 cm = ______ m

9 250 cm = ______ m

10 950 cm = ______ m

Set B

1 4 km = ______ m

2 7 km = ______ m

3 10 km = ______ m

4 $\frac{1}{2}$ km = ______ m

5 $3\frac{1}{2}$ km = ______ m

6 3000 m = ______ km

7 6000 m = ______ km

8 10 000 m = ______ km

9 500 m = ______ km

10 2500 m = ______ km

Set E

Should you use mm, cm, m or km to measure the length of:

1 a room? ______________

2 a book? ______________

3 a shoe? ______________

4 a little fingernail? ______________

5 a highway? ______________

6 a person? ______________

7 a tissue box? ______________

8 a country? ______________

9 an ant? ______________

10 a cricket pitch? ______________

Set C

1 2 cm = ______ mm

2 5 cm = ______ mm

3 10 cm = ______ mm

4 $\frac{1}{2}$ cm = ______ mm

5 7 $\frac{1}{2}$ cm = ______ mm

6 60 mm = ______ cm

7 90 mm = ______ cm

8 40 mm = ______ cm

9 5 mm = ______ cm

10 15 mm = ______ cm

Set D

1 10 mm + 10 mm = ______ cm

2 20 mm + 20 mm = ______ cm

3 30 mm + 20 mm = ______ cm

4 5 mm + 5 mm = ______ cm

5 5 mm + 15 mm = ______ cm

6 1 cm + 1 cm = ______ mm

7 2 cm + 2 cm = ______ mm

8 5 cm + 5 cm = ______ mm

9 $\frac{1}{2}$ cm + $\frac{1}{2}$ cm = ______ mm

10 $\frac{1}{2}$ cm + $\frac{1}{2}$ cm + $\frac{1}{2}$ cm = ______ mm

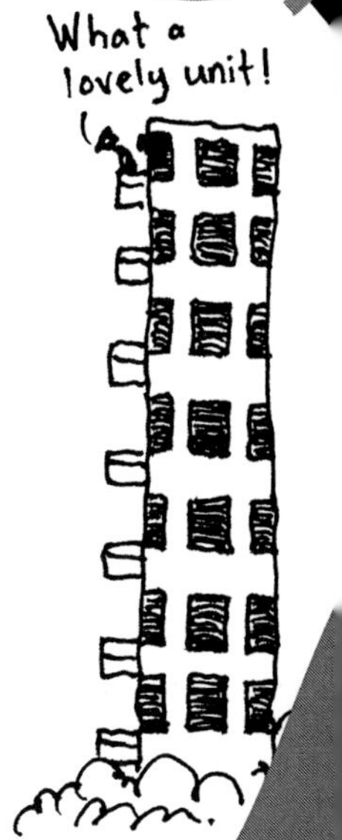

Set F

1 My book is 230 mm long. This is equal to ____ cm.

2 I rode 7000 m or ____ km.

3 My bedroom is 4000 mm long or ____ m long.

4 The city is 8 km or ________ m away.

5 My 10 cm line was only 85 mm long. It was ____ mm too short.

6 A doorway is 2 m high. Dad is 195 cm tall. His head is ____ cm below the doorway.

7 My shoe is 25 cm long and my foot is 240 mm long. The difference is ____ mm.

8 The 50 m arc in football is 500 cm from goal. True or false? ________

9 A 20 m cricket pitch is 80 m shorter than a kilometre. True or false? ________

10 How many metre rulers, laid end to end, would equal 2 kilometres in length?

____

Measurement and Geometry

Unit 23
Perimeter: Concept and missing sides

Remember that the outside, or boundary, of a shape or area is its perimeter. Shapes with similar side lengths can have their perimeters calculated easily by using multiplication.

Set A

Find the perimeters of squares with side lengths of:

1 **4 cm** _____
2 **11 cm** _____
3 **25 cm** _____
4 **7 mm** _____
5 **12 mm** _____
6 **5 m** _____
7 **8 m** _____
8 **9 km** _____
9 **15 km** _____
10 **20 km** _____

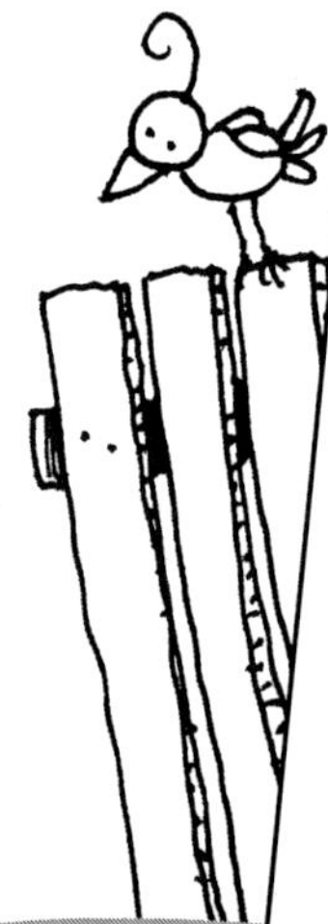

Set B

What would the side lengths of a square be if its perimeter was:

1 **12 mm?** _____
2 **16 mm?** _____
3 **24 mm?** _____
4 **8 cm?** _____
5 **28 cm?** _____
6 **44 cm?** _____
7 **20 m?** _____
8 **48 m?** _____
9 **400 km?** _____
10 **848 km?** _____

Set E

With side lengths of 5 cm, what would the perimeters of the following shapes be?

1 equilateral triangle _______
2 square _______
3 pentagon _______
4 hexagon _______
5 heptagon _______
6 octagon _______
7 nonagon _______
8 decagon _______
9 2 hexagons _______
10 2 decagons _______

Set C

What would the perimeter of a rectangle be if its length and width was:

1 6 mm and 4 mm? ______

2 12 mm and 4 mm? ______

3 8 mm and 6 mm? ______

4 10 cm and 3 cm? ______

5 20 cm and 11 cm? ______

6 15 cm and 5 cm? ______

7 3 m and 2 m? ______

8 11 m and 9 m? ______

9 2 km and 1 km? ______

10 5 km and 4 km? ______

Set D

A rectangle with a perimeter of 44 cm could have 10 different whole-number lengths and widths. How many can you find?

1 **L = ___ cm; W = ___ cm**

2 **L = ___ cm; W = ___ cm**

3 **L = ___ cm; W = ___ cm**

4 **L = ___ cm; W = ___ cm**

5 **L = ___ cm; W = ___ cm**

6 **L = ___ cm; W = ___ cm**

7 **L = ___ cm; W = ___ cm**

8 **L = ___ cm; W = ___ cm**

9 **L = ___ cm; W = ___ cm**

10 **L = ___ cm; W = ___ cm**

Set F

1 I rode 8 km around our square farm. The farm is ____ km in length and width.

2 Would the perimeter of an exercise book be 10 cm, 100 cm or 1000 cm? ______

3 Would a classroom's perimeter be 30 cm, 30 m or 30 km? ________

4 The perimeter of a house is measured in ________.

5 The perimeter of Australia is its coastline. True or false? ________

6 A shape has to be closed to have a perimeter. True or false? ________

7 Perimeter means the space inside a shape. True or false? ________

8 The perimeter of a sporting field should be measured in ________

9 You can multiply by 5 to work out the perimeter of a pentagon. True or false? ________

10 You can measure the perimeter of an octagon by multiplying by 10. True or false? ________

Measurement and Geometry

Unit 24
Area: Concept and application

Area deals with the space inside a boundary or perimeter. Remember that the area of a rectangle is worked out by multiplying together the length and the width. This divides the shape into squares such as square centimetres or square metres.

Set A

What would the areas of the following rectangles be?

1 **L = 3 cm; W = 2 cm** ________

2 **L= 10 cm; W = 5 cm** ________

3 **L = 8 cm; W = 4 cm** ________

4 **L = 12 cm; W = 6 cm** ________

5 **L = 7 cm; W = 5 cm** ________

6 **L = 11 cm; W = 8 cm** ________

7 **L = 9 cm; W = 8 cm** ________

8 **L = 10 cm; W = 8 cm** ________

9 **L = 20 cm; W = 3 cm** ________

10 **L = 100 cm; W = 10 cm** ________

Set B

Find the areas of the following squares.

1 **L = 12 m** ________

2 **L = 4 m** ________

3 **L = 7 m** ________

4 **L = 9 m** ________

5 **L = 11 m** ________

6 **L = 3 m** ________

7 **L = 8 m** ________

8 **L = 10 m** ________

9 **L = 6 m** ________

10 **L = 5 m** ________

Set C

Find the widths of the following rectangles.

1 Length = 6 cm. Area = 30 square cm. Width = ______

2 Length = 10 cm. Area = 90 square centimetres. Width = ______

3 Length = 8 cm. Area = 16 square centimetres. Width = ______

4 Length = 3 cm. Area = 6 square centimetres. Width = ______

5 Length = 11 cm. Area = 44 square centimetres. Width = ______

6 Length = 7 m. Area = 42 square metres. Width = ______

7 Length = 9 m. Area = 54 square metres. Width = ______

8 Length = 5 m. Area = 20 square metres. Width = ______

9 Length = 12 m. Area = 132 square metres. Width = ______

10 Length = 20 m. Area = 200 square metres. Width = ______

Set D

Would you use square centimetres or square metres to measure the areas of:

1 a room? ______
2 a vegetable patch? ______
3 the front cover of a book? ______
4 a stamp? ______
5 a piece of A4 paper? ______
6 a photograph? ______
7 a TV screen? ______
8 a corridor? ______
9 a wall? ______
10 a portrait? ______

Set E

Would you measure the areas of the following things in square centimetres or square metres?

1 a garage floor ______
2 a pool cover ______
3 a lawn ______
4 a welcome mat ______
5 a slice of ham ______
6 a certificate ______
7 a whiteboard ______
8 a $100 note ______
9 a footy field ______
10 a netball court ______

Set F

1 Which Australian state has the biggest area? ____________________
2 Which Australian state has the smallest area? ____________
3 The area of a rectangle equals length + width. True or false? ________
4 My classroom is bigger than 10 square metres. True or false? ________
5 The area of a computer screen is less than 1 square metre. True or false? ________
6 My bed is bigger than 1 square metre in size. True or false? ________
7 A postage stamp is about 1 square centimetre in area. True or false. ________
8 Does painting a wall deal with area or perimeter? ________
9 A 6 metre long by 5 metre wide room has an area of 22 square metres. True or false? ________
10 Cutting lawns deals with area. True or false? ________

Measurement and Geometry

Unit 25
Mass: Units and conversions

Remember that 1000 grams equals 1 kilogram and that g and kg are the abbreviations for these units of measurement.

Set A

Change into kilograms.

1 5000 g ______
2 3000 g ______
3 9000 g ______
4 11 000 g ______
5 20 000 g ______
6 500 g ______
7 1500 g ______
8 4500 g ______
9 7500 g ______
10 10 500 g ______

Set B

Change into grams.

1 3 kg ______
2 4 kg ______
3 8 kg ______
4 12 kg ______
5 30 kg ______
6 $\frac{1}{2}$ kg ______
7 $2\frac{1}{2}$ kg ______
8 $4\frac{1}{2}$ kg ______
9 $6\frac{1}{2}$ kg ______
10 $8\frac{1}{2}$ kg ______

Set E

Is it better to measure these things in grams or in kilograms?

1 a baby ______
2 a mouse ______
3 a Year 4 child ______
4 a box of biscuits ______
5 a horse ______
6 a jockey ______
7 a dollar coin ______
8 a button ______
9 a full school bag ______
10 a tennis racquet ______

Set C

How far away from 1 kilogram are the following masses?

1 **900 g** ______
2 **100 g** ______
3 **500 g** ______
4 **50 g** ______
5 **250 g** ______
6 **850 g** ______
7 **975 g** ______
8 **20 g** ______
9 **350 g** ______
10 **885 g** ______

Set D

Choose the better measure.

1 **apple: 2 g or 200 g** _____
2 **butter: 250 g or 2 kg** _____
3 **potato: 300 g or 3 kg** _____
4 **brick: 60 g or 6 kg** _____
5 **pencil: 15 g or 150 g** _____
6 **shoes: 10 kg or 1 kg** _____
7 **ball: 200 g or 2 kg** _____
8 **computer: 20 g or 2 kg** _____
9 **dog: 50 g or 5 kg** _____
10 **bottle of water: 750 g or 75 g** _____

Set F

1 I weigh 3 kilograms or _______ grams more than my sister.
2 A shoe weighs 500g. ____ shoes weigh 1 kg.
3 ____ icy poles, each weighing 100 g, will equal 1 kg.
4 How many 200 g apples will weigh 1 kg? ____
5 1/4 of a kilogram equals 200 g. True or false? ________
6 Rulers are to length as ________ are to mass.
7 Metre is to length as ________ is to mass.
8 A peach weighs 350 g. About ____ peaches will weigh 1 kg.
9 Oil floats on water. Is it heavier or lighter than water? ___________
10 If you drink a glass of water, your weight will increase. True or false? ________

Measurement and Geometry

Unit 26

Capacity/volume: Units and conversions

Remember that 1000 millilitres (mL) equals 1 litre (L) and that a litre of water weighs 1 kilogram or 1000 grams.

Set A

Change these capacities into litres.

1 **2000 mL** ______

2 **4000 mL** ______

3 **7000 mL** ______

4 **1000 mL** ______

5 **10 000 mL** ______

6 **1500 mL** ______

7 **3500 mL** ______

8 **8500 mL** ______

9 **5500 mL** ______

10 **10 500 mL** ______

Set B

Change these capacities into millilitres.

1 **3 L** ______

2 **6 L** ______

3 **8 L** ______

4 **5 L** ______

5 **9 L** ______

6 **$2\frac{1}{2}$ L** ______

7 **$4\frac{1}{2}$ L** ______

8 **$9\frac{1}{2}$ L** ______

9 **$7\frac{1}{2}$ L** ______

10 **$6\frac{1}{2}$ L** ______

Set E

How much would the following volumes of water weigh?

1 1 litre ______

2 $\frac{1}{2}$ litre ______

3 $1\frac{1}{2}$ L ______

4 4 L ______

5 $4\frac{1}{2}$ L ______

6 8 L ______

7 10 L ______

8 100 L ______

9 1000 L ______

10 10 000 L ______

Set C

Choose the more sensible measure.

1 a glass of milk: 200 mL or 2 L _______

2 a bottle of wine: 750 mL or 7 L _______

3 a bucket: 80 mL or 8 L _______

4 an egg cup: 50 mL or 500 mL _______

5 a bottle of milk: 200 mL or 2 L _______

6 a washing machine: 4 L or 40 L _______

7 a spa: 20 L or 2000 L _______

8 a bowl: 300 mL or 3 L _______

9 a vase: $1\frac{1}{2}$ L or 15 L _______

10 a petrol tank: 50 L or 500 L _______

Set D

Would you measure these capacities in millilitres or in litres?

1 a can of soup ____

2 a cup of tea ____

3 a sink ____

4 a pool ____

5 a fish tank ____

6 a dam ____

7 a drink bottle ____

8 a sauce bottle ____

9 a baby bath ____

10 a soy sauce bottle ____

Set F

1 I weighed 35 kg and then drank a litre of water. I now weigh ______

2 750 mL of water weighs ______ g.

3 A box juice holds 250 mL. ____ juice boxes will equal a litre.

4 A 200 gram bucket, containing 8 litres of water, weighs ______ grams in total.

5 Half a litre of water weighs ______ grams.

6 I lost 2 kilograms in fluids in the race. My weight dropped by ______ g.

7 A 750 mL bottle of orange juice is ______ mL short of a litre.

8 A 50 gram can, when full of soup, weighs 450 g. It holds about ______ mL of soup.

9 A cup of coffee contains 320 mL. About ____ cups equals 1 litre.

10 20 L of water comes from my hose in a minute. This is equal to ______ kg of water.

NSW: Measurement MS2.3; Qld: Measurement M3.1;
Vic: Measuring and estimating 3.3; WA: Measurement M3.1; SA: Measurement 2.5

Unit 27
Time: Digital and analogue conversions

We use both digital and analogue (or sweep-hand) clocks in our everyday lives. The following questions deal with the relationships between these two ways of telling time. Remember that there are 60 minutes in an hour—not 100!

Set A

Change these digital times to sweep-hand times.

1 3:25 ______
2 10:09 ______
3 5:22 ______
4 2:17 ______
5 8:14 ______
6 12:27 ______
7 6:04 ______
8 1:21 ______
9 4:19 ______
10 7:01 ______

Set D

Change these sweep-hand times to digital times.

1	10 minutes to 6	_____
2	24 minutes to 4	_____
3	8 minutes to 12	_____
4	20 minutes to 5	_____
5	2 minutes to 3	_____
6	7 minutes to 11	_____
7	1 minute to 2	_____
8	14 minutes to 7	_____
9	28 minutes to 1	_____
10	29 minutes to 8	_____

Set E

How many minutes from:

1	3:05 to 25 minutes past 3?	______
2	2:40 to 10 minutes to 3?	______
3	12:35 to 5 minutes to 1?	______
4	6:05 to half past 6?	______
5	7:15 to 28 minutes past 7?	______
6	20 minutes to 7 to 7:05?	______
7	12 minutes to 6 to 6:01?	______
8	3 minutes to 11 to 11:11?	______
9	4 minutes to 9 to 9:05?	______
10	6 minutes to 8 to 8:08?	______

Set B

Change these sweep-hand times to digital times.

1. 23 minutes past 8 ☐
2. 17 minutes past 11 ☐
3. 7 minutes past 6 ☐
4. 12 minutes past 4 ☐
5. 28 minutes past 10 ☐
6. 5 minutes past 7 ☐
7. 19 minutes past 2 ☐
8. 22 minutes past 12 ☐
9. 14 minutes past 3 ☐
10. 29 minutes past 8 ☐

Set C

Change these digital times into sweep-hand times.

1. **6:55** ______________________
2. **8:40** ______________________
3. **11:48** ______________________
4. **4:35** ______________________
5. **2:59** ______________________
6. **5:31** ______________________
7. **12:49** ______________________
8. **1:46** ______________________
9. **3:57** ______________________
10. **8:33** ______________________

Set F

1 I started running at 11:05 and finished at half past 11. I ran for _______ minutes.

2 My favourite half-hour TV show started at 4:35. It finished at _______.

3 The first quarter started at 2:10 and lasted for 30 minutes. It ended at _______.

4 I read for a quarter of an hour, starting at 7:50. I finished at _______.

5 Adelaide is half an hour behind Sydney. In Sydney it's 12:15. In Adelaide it's _______.

6 20 minutes ago it was 10 to 3. It's now __________________.

7 The marathon lasted for 2 hours and 10 minutes, starting at 10:00 and finishing at _______.

8 My 5-minute shower ended at 8:03. It started at _______.

9 Lunch starts at 12:40 and lasts for 55 minutes. Lunch ends at _______.

10 Our clock shows 6:11, but is 12 minutes fast. The actual time is _______.

Unit 28 Time: Unit equivalences

Time can be split up into units as small as a second and as large as a millennium. See how well you can move from one unit to another in these exercises!

Set A

1 2 minutes = _____ seconds

2 240 seconds = _____ minutes

3 3 hours = _____ minutes

4 1 1/2 hours = _____ minutes

5 12 hours is what fraction of a day? _____

6 6 hours is what fraction of a day? _____

7 2 days = _____ hours

8 240 hours = _____ days

9 100 days = _____ hours

10 1 1/2 days = _____ hours

Set D

In which seasons are the following months?

1 March __________

2 July __________

3 December __________

4 September __________

5 June __________

6 April __________

7 February __________

8 November __________

9 August __________

10 May __________

Set E

In which months do the following special days occur?

1 Christmas __________

2 Halloween __________

3 Easter __________ or __________

4 Mother's Day __________

5 Father's Day __________

6 Boxing Day __________

7 St Valentine's Day __________

8 New Year's Day __________

9 New Year's Eve __________

10 Australia Day __________

Set B

1 14 days = _____ weeks

2 5 weeks = _____ days

3 100 weeks = _____ days

4 77 days = _____ weeks

5 28 days = _____ fortnights

6 10 fortnights = _____ weeks

7 2 years = _____ weeks

8 10 years = _____ weeks

9 24 months = _____ years

10 10 years = _____ months

Set C

How many days are in the following months?

1 November: ____

2 April: ____

3 July: ____

4 December: ____

5 March: ____

6 May: ____

7 January: ____

8 October: ____

9 February: ____ or ____

10 June: ____

Set F

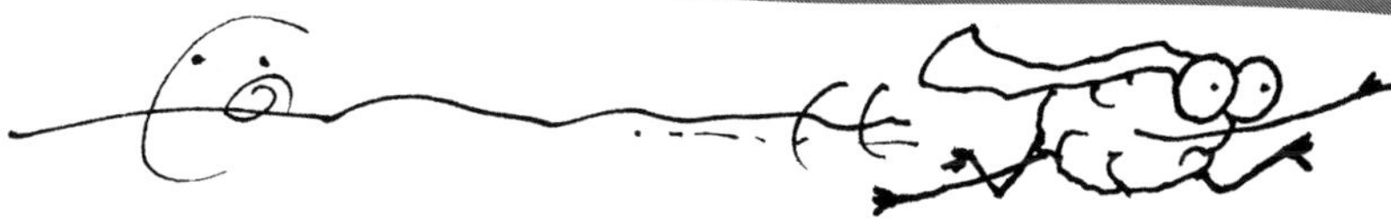

1 The marathon lasted for 2 hours and 10 minutes, or _____ minutes.

2 Which is the only month that can last for exactly 2 fortnights?

____

3 I am 10 and so I must have lived for _____ months.

4 A 1 1/2-hour movie lasts for _____ minutes.

5 Spring contains the months of _______________, _______________ and _______________.

6 The last month of summer is _______________.

7 I earn $10 a week in pocket money. In a year I earn _____.

8 Year 1 to Year 4 consists of _____ terms.

9 The 40-hour famine lasts for 1 day and _____ hours.

10 I sleep for 8 hours a night. In _____ days this will equal a whole day of sleep.

Measurement and Geometry

Unit 29
Money: Economical ways of making amounts

Making amounts of money economically means as quickly as possible by using the smallest number of notes and coins as is possible. Remember to always start with the most valuable note or coin.

Set A

How many coins would be required to make the following amounts as quickly as possible?

1. 25 cents ☐
2. 30 cents ☐
3. 45 cents ☐
4. 50 cents ☐
5. 60 cents ☐
6. 80 cents ☐
7. 85 cents ☐
8. 90 cents ☐
9. 95 cents ☐
10. $1.00 ☐

Set B

How many coins would be required to make the following amounts as quickly as possible?

1. $1.20 ☐
2. $2.40 ☐
3. $1.80 ☐
4. $2.05 ☐
5. $2.55 ☐
6. $3.10 ☐
7. $3.80 ☐
8. $4.00 ☐
9. $4.50 ☐
10. $4.95 ☐

Set E

How many notes and coins would be required to make the following amounts of money as quickly as possible?

1. $6.20 ____
2. $6.40 ____
3. $7.80 ____
4. $8.40 ____
5. $9.20 ____
6. $9.95 ____
7. $10.30 ____
8. $12.10 ____
9. $16.50 ____
10. $18.40 ____

Set C

How many notes would be required to make the following amounts as quickly as possible?

1	$30	☐
2	$45	☐
3	$50	☐
4	$60	☐
5	$65	☐
6	$70	☐
7	$80	☐
8	$85	☐
9	$90	☐
10	$95	☐

Set D

How many notes would be required to make the following amounts as quickly as possible?

1	$105	☐
2	$110	☐
3	$140	☐
4	$180	☐
5	$195	☐
6	$205	☐
7	$280	☐
8	$310	☐
9	$440	☐
10	$565	☐

Set F

1. I paid $7.50 for a cinema ticket and paid with $10. My economical change was in ____ coins.
2. My friend bought his ticket and paid with a $20 note. His change was $________.
3. Dad paid $43 for the football as economically as possible. He used ___ notes and ___ coins.
4. What change from $100 would you get if you bought a shirt for $79.95? _______
5. How should the change from the shirt be given? _____________
6. How much money would you have if you had one of every silver coin? ____________
7. How much money would you have if you had one of every type of coin? _______
8. How much money would you have if you had one of every note? _______
9. Make $85.00 as quickly as possible. ______________________
10. Make $245.75 as quickly as possible. ____

Measurement and Geometry

Unit 30
2D and 3D: Properties and directions

2D shapes, called polygons, are flat and can be drawn on a piece of paper. Polygon comes from Greek ('many corners').

3D shapes, called polyhedra, have depth and can be picked up. Polyhedra comes from Greek ('many faces').

North, south, east and west are the four main compass points.

Set A

How many sides are on the following polygons?

1 2 triangles ☐
2 3 rectangles ☐
3 3 rhombuses ☐
4 5 squares ☐
5 5 pentagons ☐
6 2 hexagons ☐
7 3 heptagons ☐
8 4 octagons ☐
9 2 nonagons ☐
10 100 decagons ☐

Set B

How many faces are on the following polyhedra?

1 cube ☐
2 rectangular prism ☐
3 cylinder ☐
4 triangular prism ☐
5 hexagonal prism ☐
6 sphere ☐
7 square pyramid ☐
8 hexagonal pyramid ☐
9 cone ☐
10 octagonal pyramid ☐

Set E

Start by facing north. Which direction would you be facing if you went:

1 a quarter turn right? ________
2 a quarter turn left? ________
3 a half turn right? ________
4 a half turn left? ________
5 a three-quarter turn right? ________
6 a three-quarter turn left? ________
7 a full turn right? ________
8 a full turn left? ________
9 1 1/2 turns right? ________
10 1 1/2 turns left? ________

Set C

How many edges are on the following polyhedra?

1 cube ☐
2 rectangular prism ☐
3 cylinder ☐
4 triangular prism ☐
5 hexagonal prism ☐
6 sphere ☐
7 square pyramid ☐
8 hexagonal pyramid ☐
9 cone ☐
10 octagonal pyramid ☐

Set D

How many vertices or corners are on the following polyhedra?

1 cube ☐
2 rectangular prism ☐
3 cylinder ☐
4 triangular prism ☐
5 hexagonal prism ☐
6 sphere ☐
7 square pyramid ☐
8 hexagonal pyramid ☐
9 cone ☐
10 octagonal pyramid ☐

Set F

1 A rectangle can tessellate. True or false? ________

2 A circle can tessellate. True or false? ________

3 A hexagon can tessellate. True or false? ________

4 Which two shapes are used to make a rectangular prism?

____

5 Which two shapes are used to make a triangular prism?

____

6 North is opposite of east. True or false? ________

7 I have 7 similar shapes that have 42 sides altogether. They are ________________.

8 I have 9 similar shapes that have 72 sides altogether. They are ________________.

9 A tube is an example of a ________________.

10 Three cubes stuck end to end make a ______________________________.

NSW: Space and geometry SGS2.1, 2.2a; Qld: Space S3.1, 3.2; Vic: Shape and space 3.2, Location 3.1; WA: Space S3.2, 3.1; SA: Spatial sense and geometric reasoning 2.12, 2.14

Statistics and Probability

Unit 31 Chance: Concepts of likelihood

For these questions, imagine that you have 10 cards numbered 1 to 10. Cards 1 to 5 are red and cards 6 to 10 are black. Shuffle the cards and have them face down.

Set A

How many cards are:

1 even? ☐
2 odd? ☐
3 red *or* black? ☐
4 bigger than 2? ☐
5 smaller than 5? ☐
6 between 3 and 8 inclusive? ☐
7 bigger than a half of 12? ☐
8 equal to 3 x 3? ☐
9 red and even? ☐
10 black and odd? ☐

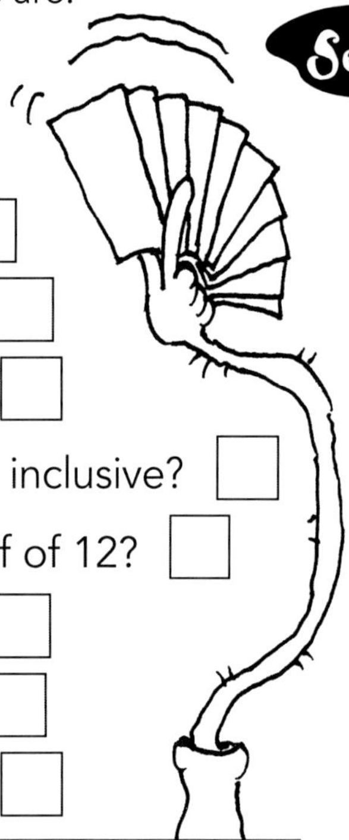

Set B

What are the chances of drawing the following cards? Write 'certain', '50/50', or 'impossible'.

1 a 9 ___________
2 an even-numbered card ___________
3 an 11 ___________
4 a 2 or a 3 ___________
5 a number less than 20 ___________
6 a red card ___________
7 a black card ___________
8 either a red or a black card ___________
9 a green card ___________
10 a 3, 4, 5, 6 or 7 ___________

Set E

Answer true (T) or false (F) to the following statements.

1 The chance of drawing a red card is the same as for a black card. ____
2 The chance of drawing an even card is the same as for a black card. ____
3 The chance of drawing an odd card is the same as for a black card. ____
4 The chance of drawing an odd card is the same as for a red even card. ____
5 The chance of drawing an odd card is the same as for a red odd card. ____
6 The chance of drawing an odd card is the same as for a black even card. ____
7 The chance of drawing a 1, 3, 4, 7 or 9 card is the same as for a black card. ____
8 The chance of drawing a 1 or 9 card is the same as for a 2 or 10 card. ____
9 A 9 or 10 card has a greater chance of being drawn than a 1 or 2 card. ____
10 The chance of drawing a 5 card is only half the chance of drawing a 10 card. ____

Set C

Circle the card that has the better chance of being drawn. If it is a 50/50 chance, don't circle either.

1 A card less than 7 or an odd number

2 An odd card or a red card

3 An even card or an odd card

4 A 6 or a 10

5 A card greater than 4 or a card smaller than 4

6 A card greater than 6 or a card smaller than 4

7 A card with 3 letters in its number name or an even card

8 A card greater than 1 or a card smaller than 8

9 An odd red card or an odd black card

10 An even red card or an odd black card

Set D

What is the chance of drawing (answer as a fraction):

1 a 6? ☐

2 a 10? ☐

3 an even number? ☐

4 an odd number? ☐

5 a 5 or 6? ☐

6 a 7, 8 or 9? ☐

7 a card bigger than 7? ☐

8 an even red card? ☐

9 an odd black card? ☐

10 a card smaller than 10? ☐

Set F

1 How many of the 10 cards have 3 letters in their number name? _____

2 What is the chance of drawing one of these cards? _____

3 How many of the 10 cards have more than 3 letters in their number name? _____

4 What is the chance of drawing one of these cards? _____

5 How many cards are red and have 3 letters in their number name? _____

6 What is the chance of drawing one of these cards? _____

7 How many of the black cards have more than 3 letters in their number name? _____

8 What is the chance of drawing one of these cards? _____

9 How many cards greater than 6 have 4 letters in their number name? _____

10 What is the chance of drawing such a card? _____

Unit 32 Revision: All sorts

Revision

Here are some revision questions that cover all of the things you have worked through since Unit 1.

Good luck!

Set A

1 20 = 6 + ___

2 20 – 11 = ___

3 Double 23 ___

4 42 = ___ x 7

5 99 divided by 11 ___

6 The 12th odd number ___

7 Ten thousand has ___ zeroes.

8 126 to the nearest ten _____

9 12, 21, 30, ___

10 3 wholes equal ___ halves.

Set B

1 3 m = _____ cm

2 12 m x 4 = ___

3 2 kg of water = ___ L

4 10:45 to 11:05 = ___ minutes

5 $4.40 is made with ___ coins

6 5 nonagons have ___ sides

7 Write the fraction one-eighth ___

8 34 + 30 = ___

9 56 – 20 = ___

10 8 x 8 = ___

Set E

1 Round 1744 to the nearest 100 ______

2 3, 4, 6, 9, ______

3 0.7 equals the fraction ______

4 3 km = ______ m

5 A pentagon with side lengths of 7 cm will have a perimeter of ______ cm.

6 5 L = ______ mL

7 What day of the week will it be 5 weeks after Friday? ______

8 How many coins will equal $4.70 made as quickly as possible? ______

9 A half turn from west will point to ______

10 As a fraction, the chance of cutting a 5 with cards numbered 1 to 6 is ____

Set C

1 $50 \div 7 = 7$ r ___

2 7 + 7 doubled ___

3 $24 \div 2 \div 2 =$ ___

4 1 000 = 100 x ___

5 2, 4, 8, ___, 32

6 $4 \times 4 \div 2 - 5 =$ ___

7 4 m x 10 m = ___ square m

8 12 000 g = ___ kg

9 May is in the season of __________

10 4 weeks = ___ days

Set D

1 \$36 = \$20 + ___ + ___ + ___

2 A basketball's shape ________

3 West is opposite ________

4 Even money is a ___/___ chance.

5 27 + 7 = ___

6 74 – 5 = ___

7 12 x 6 = ___

8 63 divided by 9 = ___

9 3 x 4 added to 11 = ___

10 The 100th even number ___

Set F

1 Find the sum of 11 and 12. _____

2 Find the difference between 20 and 8. _____

3 The product of 6 and 12 is _____

4 27 cakes shared equally among 9 children means that they will receive _____ cakes each.

5 Which number, when halved, equals 54? _____

6 Ten tenths equals ________________________.

7 \$84 made as economically as possible, will need _____ notes and _____ coins.

8 June has _____ days and is in the season of ____________.

9 2 cubes have _____ edges altogether.

10 With 10 blue cards and 5 red cards, the chance of picking a red card, as a fraction, is _____.

Glossary

Add	To group together
Altogether	The answer to an addition problem
Analogue clock	A sweep-hand clock
Autumn	March, April, May
Bi	A prefix meaning 2
Centi	A prefix meaning 1/100
Centimetre	100th of a metre
Century	100 years
Composite	A number with more than two factors
Day	24 hours
Deca	A prefix meaning 10
Decade	A period of 10 years
Decagon	A 10-sided shape
Difference	How far one number is away from another number
Divide	To split up into equal pieces or to share
Dodecagon	A shape with 12 sides
Double	To add a number onto itself
Dozen	12
Edge	A boundary
Equilateral	Sides the same length
Face	A side of a 3D shape
Factor	What divides into a number
Fortnight	14 days
Gram	The standard unit for mass
Groups of	Counting in lots
Halve	Divide into two equal parts
Hendecagon	A shape with 11 sides
Heptagon	A shape with 7 sides
Hexagon	A shape with 6 sides
Hour	60 minutes
Kilo	A prefix meaning 1000
Kilogram	1000 grams
Kilometre	1000 metres
Leap year	A year of 366 days
Litre	The standard unit of capacity
Margin	How far two numbers are apart
Metre	The standard unit of length
Millennium	1000 years
Millilitre	One-thousandth of a litre
Minute	One-sixtieth of an hour
Month	One-twelfth of a year
Multiple	The answers to times tables
Multiply	Count groups
Nonagon	A shape with 9 sides
Octa	A prefix meaning 8
Octagon	A shape with 8 sides
Penta	A prefix meaning 5
Pentagon	A polygon with 5 sides
Perimeter	The length of the boundary of a shape
Polygon	A 2D shape with many corners
Polyhedra	A 3D shape with many faces
Prime	A number with just two factors
Product	The answer to a multiplication problem
Quad	A prefix meaning 4
Quadrilateral	A 4-sided shape
Quadruple	To multiply by 4
Quotient	The result of a division question
Regular (shape)	Having the same side lengths
Remainder	What is left over after a division sum has been completed
Remove	To take away
Rhombus	A 4-sided figure with equal sides but no right angles
Right angle	An L-shaped angle of 90 degrees
Round	To take a number to the nearest 10, 100, 1000 etc.
Rule	The pattern in a sequence
Season	A quarter of a year
Second	One-sixtieth of a minute
Semester	Half a year at school
Sequence	A number pattern
Share	To divide into equal pieces
Solve	To find the answer to a question
Spring	September, October, November
Subtract	To take away
Sum	The answer to an addition problem
Summer	December, January, February
Term	Each member of a sequence
Term	One quarter of a year at school
Tessellate	To fit together without leaving gaps
Times	Groups of
Total	The answer to an addition problem
Tri	A prefix meaning 3
Triple	To multiply by 3
Twice	2 times, or to double
Vertex	The corner where lines meet on 2D or 3D shapes
Week	7 days
Winter	June, July, August
Year	365 days